Sheet Metal Forming Optimization

Bioinspired Approaches

Sheet Metal Forming Optimization

Bioinspired Approaches

Ganesh M. Kakandikar and Vilas M. Nandedkar

CRC Press
Taylor & Francis Group
Boca Raton London New York

CRC Press is an imprint of the
Taylor & Francis Group, an **informa** business

CRC Press
Taylor & Francis Group
6000 Broken Sound Parkway NW, Suite 300
Boca Raton, FL 33487-2742

First issued in paperback 2020

ISBN 13: 978-0-367-57286-0 (pbk)
ISBN 13: 978-1-4987-9614-9 (hbk)

Library of Congress Cataloging-in-Publication Data

Names: Kakandikar, Ganesh M., author. | Nandedkar, Vilas M., author.
Title: Sheet metal forming optimization : bioinspired approaches / Ganesh M. Kakandikar, Vilas M. Nandedkar.
Description: Boca Raton : Taylor & Francis, a CRC title, part of the Taylor & Francis imprint, a member of the Taylor & Francis Group, the academic division of T&F Informa, plc, [2018] | Includes bibliographical references and index.
Identifiers: LCCN 2017020157| ISBN 9781498796149 (hardback : alk. paper) | ISBN 9781315156101 (ebook)
Subjects: LCSH: Sheet-metal work.
Classification: LCC TS250 .K35 2018 | DDC 671.8/23--dc23
LC record available at https://lccn.loc.gov/2017020157

Visit the Taylor & Francis Web site at
http://www.taylorandfrancis.com

and the CRC Press Web site at
http://www.crcpress.com

Contents

Preface

Sheet metal forming has been extensively used in manufacturing domestic and industrial products. Metal forming involves a wide range of operations from simple punching to complicated deep drawing. The recent developments in the automotive sector and aerospace industry have attracted many researchers toward metal forming studies. The application of newer materials, tailor-welded blanks, and process optimization are areas of interest. There is a great demand and pressure on the industry to produce reliable parts with lower weight and higher strength.

Metal forming is a complicated process to understand and operate. It involves many parameters that are strongly interdependent. These include material characteristics, machine parameters, and process parameters and also the work piece geometry that has to be manufactured. An extensive list of such parameters will include material thickness, work hardening coefficient N, plastic strain ratio R, blank size and shape, punch nose radii, draw ratio, die profile radii, lubricant and friction condition, draw bead height and shape, blank-holder force, and forming/punch load.

Many evolutionary and bioinspired algorithms have attracted several researchers to apply them in engineering applications for optimization. They offer greater advantages of applicability and computational intelligence. Thanks to the excellent results obtained, they are becoming increasingly popular.

This book provides an integrative study and optimization of metal forming, especially drawing/deep drawing. Chapter 1 presents the basics of metal forming, formability. Chapter 2 discusses the process parameters in detail. Chapter 3 extensively analyzes prominent modes of failure. Chapter 4 describes the basics of optimization and various bioinspired approaches in brief. Chapters 5 through 11 present optimization studies on various industrial components applying bioinspired optimization algorithms. The authors have developed and applied a novel approach combining finite element simulation, Taguchi design of experiments, and bioinspired optimization. This novel approach has been validated using experimentation. The authors are confident that the book will serve as a helpful companion for researchers working on other metal-forming processes. This has also validated the application of bioinspired approaches to metal forming.

Authors

Dr. Ganesh M. Kakandikar is presently working as professor in the Department of Mechanical Engineering at MAEER's Maharashtra Institute of Technology, Kothrud, Pune, India. He completed his PhD from Swami Ramanand Teerth Marathwada University, Nanded. He has 19 years of teaching experience. His areas of expertise are sheet metal forming, optimization technique, and CAD/CAM/CAE. He has 55 national and international publications to his credit and has authored three international books.

Dr. Vilas M. Nandedkar is professor in the Department of Production Engineering at S.G.G.S. Institute of Engineering & Technology, Nanded, India. He completed his PhD from the Indian Institute of Technology, Powai, Mumbai. His areas of interest are robotics, sheet metal forming, and technology management. He has authored two international books and has 150 national and international publications to his credit.

1

Introduction to Metal Forming

1.1 Introduction

There has been tremendous development in the automotive and aerospace industry in the last decade, especially in India. The major manufacturing strategy for constructing bodies of automotive and aerospace structures is sheet metal forming. It involves a cluster of significant manufacturing processes from simple bending to deep drawing. A wide range of products from body panels, fenders, and wing parts to consumer products like kitchen sinks, cans, and boxes are made with precision through different operations on a plane blank. One of the most significant operations is the forming process, which is also termed drawing/stamping, and in some cases deep drawing, when the depth of the drawn part exceeds its diameter. This process includes a wide spectrum of operations and flow conditions. Simultaneous compression–tension can be observed in the radial and circumferential direction. The forming process involves rigid tooling, draw punches, a blank holder, and a female die. The blank is generally kept over the die and pushed by the draw punch into the die radially; pressure is applied on the blank holder to control the flow of material. Accurate control of material can avoid dominance of compression, resulting in wrinkling. With developments in the automotive and aerospace industry, the formability of metals is gaining more and more attention. This is due to the continuous demand on the industry to produce lightweight components. Materials with higher strength to weight ratios are being used in forming parts of missiles, aircraft, and, more recently, even in automobile industries. There is also significant ongoing research on materials and composites for these applications.

On the other hand, design in sheet metal forming is considered an art rather than a science in small-scale industries as it depends on the operator's knowledge, experience, and intuition. Computer-aided design and computer-aided engineering have not yet percolated to that level. This is due to the involvement of many parameters and their interdependence in the forming process. These include properties of material, punch and die geometry, work piece profile, and working conditions. The other reason is that research and development in sheet metal forming requires extensive and costly prototype testing and experimentation to arrive at a competitive product. An example is the traditional method of designing die geometry for the sheet metal stamping process. A working design is arrived at, based on the competency of the designer and by trial and error. The goals of the sheet metal forming process are to minimize the time and cost of manufacturing while optimizing the quality of the product. It is very difficult to satisfy these goals because the success of the deep drawing part depends upon the experience acquired by the tool design engineer over the years.

Thus, a large amount of time and money is utilized in an industry in finding appropriate tool geometries and manufacturing parameters by trial and error, after which they are modified by performing repeated experiments till a feasible solution is obtained. This leads to the sheet metal forming process simulation, which has emerged as the need of the hour. As sheet metals are transformed from their initial to final shape, the process involves large displacements and induced strains. Finite element methods play a very important role in reducing cost as well as time to obtain a complex physical phenomenon, and also help to better understand the process and to control the quality of the product. Thus, numerical simulation of the sheet metal forming process is a very useful tool for analysis.

1.2 Drawing/Deep Drawing Process

Drawing/forming/stamping is a process in which the form of a flat blank is changed into the desired component. Drawing is mostly applied in metal-working industries to produce cup-shaped components with high manufacturing efficiency on mechanical or hydraulic press. Cup drawing is a basic procedure to test for the sheet metal formability prior to forming other shape components. Typically in drawing, the blank is usually constrained over the draw punch inside the die to give the required shape of the cavity. Large plastic deformation combined with a complex flow of material is required for a successful process (Figure 1.1).

The physics of the process has two main issues: considerable decrease in the limiting drawing ratio and occurrence of wrinkles. Both must be controlled accurately. For controlling wrinkles in the flange and cup wall, a blank holder is utilized. A high blank holder force causes tearing by resisting the easy flow of materials while a lower one results in wrinkling. Wrinkling is primarily initiated by localized buckling, due to compressive stress in a circumferential direction. Radial tensile stress causes thinning of the material, resulting in tearing. A balance in the compression and tension of the limiting drawing ratio is important to determine the allowable strain capacity. Drawn parts vary in shape, size, and configuration. Every part has a different deforming position, deforming characteristic, strain and stress distribution such as circular/cylindrical, and

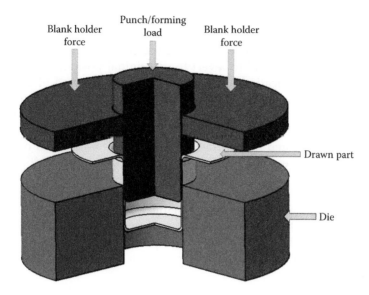

FIGURE 1.1
Deep drawing process—schematic representation.

a rectangular and irregular/nonaxis symmetry. During the process, with the progression of the punch over the blank into the die, friction is generated in the contact region. Frictional forces generated at the die shoulder and on the punch face influence the deformation of the work piece. Often additional frictional forces are introduced in the forming process (Figure 1.2).

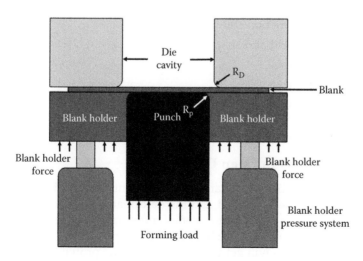

FIGURE 1.2
Deep drawing process—contact regions.

In special cases, such as the forming of large radii, shallow parts, with small diameters, additional deformation is required to achieve the desired final shape. In these situations, a springback mostly occurs. Springback is the capacity of sheet metal to revert to its original position after removal of the load. The lubrication at the interface of the die and blank, punch and blank, as well as blank holder and blank also affects the quality of the component. The friction coefficient μ is the main indicator, which is influenced by the material, the contact surface, and the lubricant. Smooth flow of material is facilitated by appropriate punch nose radius and die profile radius in tooling. Researchers have found that larger formability can be achieved when the cup is drawn with a larger die profile radius. With high strength and lesser thickness of sheet metals, the problem of fracture and wrinkling becomes prevalent. Thus, many parameters that have already been discussed and more that are listed further have a greater influence on the success of the process. These are properties of material, machine parameters such as tool and die configurations, work piece geometry, and process conditions:

- Material thickness
- Work hardening coefficient N
- Plastic strain ratio R
- Blank size and shape
- Part geometry
- Punch nose radii
- Draw ratio
- Die profile radii
- Lubricant and friction condition
- Draw bead height and shape
- Blank holder force
- Forming/punch load

1.3 State of Stress in Deep Drawing

In drawing a cup the metal can be divided into three different regions based on the stresses developed and the deformations observed. Figure 1.3 describes the various deformation zones and stresses represented in a pie-shaped section. The metal at the bottom of the cup, which is in contact with the head of the punch, will get wrapped around the profile of the punch by

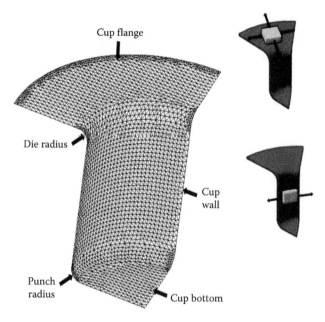

FIGURE 1.3
State of stresses in deep drawing process.

virtue of biaxial stresses. The metal in the outer portion of the blank is compressed radially inward, toward the throat of the die and flows over the die profile radius. As the material travels over the die, the outer circumference continuously decreases from that of the original blank, πd_0, to that of the finished cup, πd_1, where d_0 is the blank diameter and d_1 is the cup diameter. The material in the wall region experiences compressive strain in the circumferential direction and a tensile strain in the direction of the radius. There is continuous increase in thickness by virtue of these two principal strains, when the metal moves inward.

Bending and straightening of sheet metal happens over the die radius, while it is subjected to tensile stresses at the same time. Plastic bending under tension results in significant thinning, and this balances the thickening by virtue of circumferential shrinking. There is a narrow ring of metal in between the inner stretched and outer shrunk zone, which is bent neither on the die nor on the punch. Only simple tensile loading takes place in this region throughout the drawing operation. If the drawing ability of the sheet metal is exceeded, it initiates cracking in this region. The drawing operation exhibits positive major strain and negative minor strain during deformation.

1.4 Description and Significance of Intrinsic Sheet Metal Properties

Deformation of the sheet is influenced by material properties like strain-hardening coefficient (n), strain rate sensitivity (m), and anisotropy parameter (r).

1.4.1 Strain Hardening Coefficient "n"

The strain-hardening coefficient "n," also known as the strain-hardening index, is a material constant that indicates stress–strain behavior in work hardening. It is a quantitative measurement of the strain-hardening characteristic of a material. It can be defined as

$$n = \frac{d * \ln(\sigma_T)}{d * \ln(\varepsilon)} \tag{1.1}$$

where
  σ_T is the true stress (load/instantaneous area)
  ε is the true strain

It is the slope of the logarithmic true stress–strain curve. The strain-hardening coefficient n is determined by the dependence of the flow stress on the level of strain. The strain-hardening exponent n is a primary metal property that can be determined by a simple tension test or from the measurement of strain in a special specimen. When metal alloys are cold-worked, their yield strength increases. The n value is the amount of strengthening for each increment of straining. The higher the n value, the steeper is the stress–strain curve in the uniform elongation region of the tensile test.

The strain-hardening coefficient is determined by the dependence of the flow stress on the level of strain. In materials with high n value, the flow stress increases rapidly with strain. This results in the distribution of strain uniformly throughout the sheet, even in low-strain areas. As a consequence of uniform deformation, the forming limit or formability increases. A high n value indicates good formability in the stretching window. The constitutive equation

$$\sigma = k * \varepsilon^n \tag{1.2}$$

is a good approximation for the true stress–strain behavior of the material where k is the strength coefficient of the material. Most steels with yield strengths below 345 MPa and many aluminum alloys have an n value ranging from 0.2 to 0.3. For many higher yield strength steels, n is given by the relation n = 70/yield strength in MPa. For materials with high n values, a large difference is observed between the yield strength and the ultimate tensile strength. The ratio of these properties therefore provides another measure of formability.

1.4.2 Strain Rate Sensitivity "m"

The yield criterion, which governs the plastic flow, was assumed to be independent of the rate of strain. However, the plastic flow of some materials is dependent on the strain rate, which is known as material strain rate sensitivity, or viscoplasticity. The strain rate sensitivity of flow stress for superplastic material is a very important mechanical characteristic. Like the strain-hardening coefficient, strain rate sensitivity can be defined as

$$m = \frac{d*\ln(\sigma_T)}{d*\ln(\varepsilon_t)} \tag{1.3}$$

where σ_T is the true stress and ε is the true strain rate, and

$$\varepsilon = \frac{d\varepsilon}{dt}$$

This is the slope of the logarithmic plot of the true stress–strain rate curve. A material is usually considered to be superplastic if its m value is greater than 0.3. Superplastic alloys have low flow stresses compared with those of conventional materials. During tensile deformation, the effect of a high m is that it inhibits catastrophic necking. The m values of commercial superplastic alloys lie in the range of 0.4–0.8. Positive strain rate sensitivity means there is an increase in flow stress with the rate of deformation. This has two consequences. First, higher stress is required to form parts at higher rates. Second, at a given forming rate, the material resists further deformation in regions that are strained more rapidly than the adjacent regions by increasing the flow stress in these regions. This uniformly distributes the strain. The distribution of higher stress in a forming operation is usually not a major consideration, but the ability to distribute strain can be crucial. This becomes particularly important in the postuniform elongation region, where necking and high strain concentrations occur. Most superplastic materials have a high m value in the range of 0.2–0.7, which is one or two orders of magnitude higher than typical values for steel. Higher n and m values result in better formability in stretching operations, but have little effect on drawing ability. In a drawing operation, metal in the flange must be drawn in without causing failure in the wall. In this case, high n and m value strengthen the wall, which is beneficial, but they also strengthen the flange and make it harder to draw in, which is detrimental.

1.4.3 Normal and Planar Anisotropy

The plastic strain ratio r is defined as the ratio of the true width strain to the true thickness strain in the uniform elongation region of a tensile test (Figure 1.4):

$$r = \frac{\varepsilon w}{\varepsilon t} = \frac{(wf/wo)}{\ln(tf/to)} \tag{1.4}$$

Assume the volume remains constant:

$$\varepsilon t = \ln(lo * wo / lf * wf) \tag{1.5}$$

$$r = \frac{\ln(w/wo)}{\ln(lo * wo / lf * wf)} \tag{1.6}$$

The r value is a measure of the ability of a material to resist thinning. This value is dependent on the direction in the sheet. It is therefore common practice to measure the average r value or average normal anisotropy (R-bar) and the planar anisotropy (Δr). The property R-bar is defined as R-bar = (r_0 + 2r_{45} + r_{90})/4, where the subscripts refer to the angle between the tensile specimen axis and the rolling direction. The value Δr is defined as Δr = (r_0 – 2r_{45} + r_{90})/2 and is a measure of variation of r with the direction of the plane of a sheet.

In drawing, material in the flange is stretched in one direction (radially) and compressed in the perpendicular direction (circumferentially). A high r value indicates a material with good drawing properties. In a drawing operation involving a cylindrical cup, this variation leads to a cup with a wall that varies in height, a phenomenon known as earing. The value of R-bar determines the average depth of the deepest draw possible. The value of Δr determines the extent of earing. The combination of a high R-bar and a low Δr provides optimal drawing ability.

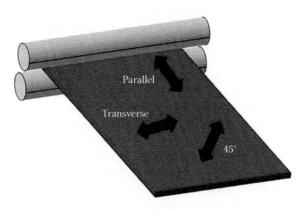

FIGURE 1.4
Rolling process.

1.5 Formability of Sheet Metal

Formability is concerned with sheet metal forming. Most sheet metal operations involve extensive tensile deformation. Localized tensile deformation due to thinning, called necking and fracture, takes place. Anisotropy plays a major role in sheet metal operations. Formability can be defined as the ease with which a sheet metal can be converted into the required shape without undergoing any localized necking, thinning, or fracture. Formability is the capability of sheet metal to undergo plastic deformation to assume a given shape without any defects. Metal-forming operations involve complex stresses from bending to deep drawing, so it is necessary to estimate how much material can be deformed safely for designing reproducible forming operations. Material properties like ultimate tensile strength, yield strength, Young's modulus, ductility, and hardness have a significant influence on formability. For plane strain deformation, the critical strain at which localized necking or plastic instability occurs can be proved to be equal to 2n, and for uniaxial tensile it is equal to n, where n is the strain-hardening exponent. Hence materials with higher n values exhibit higher necking strain, thus delaying necking considerably. Few materials exhibit diffuse necking. A simple uniaxial tensile test has limitations when dealing with formability. The biaxial or triaxial nature of stress acting on the sheet metal can be observed during forming. Many formability tests have been proposed in the literature for different classes of materials. Loading paths are also changed during metal forming due to tool geometry or metallographic texture.

1.6 Strain-Defined Failure-Limit Curves

The formability limits of sheet metal operations are described in terms of the principle strains with the forming-limit diagram (FLD). The first to attempt to develop the FLD were Keeler and Backofen, who performed tests on circular blanks with a hemispherical punch. They observed that the ratio of the circumferential strains to the radial strains, that is, the in-plane–strain ratio, increases from 0 to 1, and the fracture strain increases rapidly. Keeler employed electrochemically etched grids to measure strains and strain distributions for determining forming limits. He thus prepared a diagram with major and minor strains on both axes that represent the "critical strain level" as a curve, which increases as the strain changes to a balanced biaxial tension from the in-plane strain.

In 1967, Marciniak and Kuczynski proposed their analytical model for limit strain prediction with reference to the initial homogeneity of a material. This model, known as the M-K Theory, became the foundation for most of the work conducted in this area. They accounted for changes in limiting strain values to the loading history of the specimen, the ratio of the principal stresses, and several material properties, along with initial homogeneity. They observed that the maximum limiting strains in the biaxial stretching zone were obtained when equi-biaxial tension was applied, while the minimum values were obtained in the in-plane–strain areas. Circular grid analysis and FLD proposed by Keeler became popular methods for calculating strain distributions in actual stamped parts in order to improve the quality and optimize the die design.

The results of Keeler's experiments confirmed that the lowest formability can only be achieved under plane-strain conditions. He restricted the inward flow of metal from the flange to induce biaxial tensile strains, which increased the forming limits.

His focus was on forming limits in biaxial tension regions. However, in 1968, Goodwin proposed a grouping of cup and tension tests to define a failure band with the negative and positive quadrants of a minor strain. This primarily defined the general form of the FLD, as shown in Figure 1.5. Marciniak et al. studied the effects of planar anisotropy on the formability of steel, aluminum, and copper.

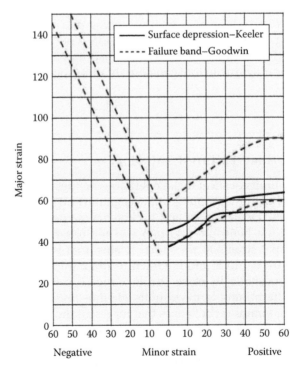

FIGURE 1.5
Keeler–Goodwin failure-limit diagram.

Proportional-straining conditions were employed for these tests. The results revealed that the limits of forming were appreciably changed from rolling to transverse direction. They concluded with the observation that limit strains are mostly higher in the rolling direction. The work was extended with variations in the FLD by Ghosh and Hecker. The outcome was that the limit strains in punch stretching were significantly higher than those from in-plane stretching.

They achieved in-plane stretching by applying a polyethylene spacer on either side of the punch nose. The forces of friction and curvature during stretching initiate fast strain localizations compared to in-plane stretching. By this time, every researcher employed the circle grid analysis to analyze the forming limits on drawn automotive stampings or on test specimens deformed with punches of various geometries. To study various strain conditions as uniaxial tension to balanced biaxial tension, Hecker, in 1975, came up with an approach of stretching sheets with various widths over a hemispherical punch.

Hecker defined a single limit of failure at the onset of localized necking, lying almost in the same band as that defined by Keeler–Goodwin. It became possible now to determine an entire forming-limit curve (FLC) with specimens of different specifications using different lubricants.

Lee and Zaverl proposed an analytical model, elaborating the growth of necking in sheet metals, under various load conditions, to plot FLDs. Their work validated the results of earlier researchers of the lower, in-plane forming limits. They could not, however, establish a relationship between the history of straining and its effect on the plastic–strain ratio. Hsu et al. conducted experiments to show that axisymmetric cup-drawing operations do not exhibit linear strain paths. There is a greater change in the strain paths when the flange is drawn inside the cup, but the initial and latter stages of deformation vary. All the research work related to FLDs was based on linear strain paths. The circle grid analysis and numerical simulations made it possible to readily understand the formability and strain history. Hecker determined the actual strain paths that occurred during a hemispherical-punch FLD test.

Soon after this, Keeler–Goodwin proposed their FLD. Researchers found that the FLD has significant effects on strain paths causing alterations in them. Matsuoka and Sudo performed experiments on aluminum-killed steel sheets, subjecting them to second-stage deformations. The first-stage deformation that was applied was either balanced biaxial tension, uniaxial tension, or a modified drawing operation. The outcome of the study was an increase in forming limits with higher first-stage strains. The first-stage biaxial tension was an exception, as it resulted in significantly lower fracture strains. Aluminum-stabilized steel sheets were investigated by Kobayashi et al. for strain paths depending on forming limits. Their findings were in agreement with Matsuoka and Sudo's work, which stated that equi-biaxial prestrains reduce the FLC.

They also agreed that predeformations with plane strain contribute to lowering of forming limits. Uniaxial prestrain tension extended the forming limits for subsequent equi-biaxial stretching to a higher extent compared to constant strain ratio deformation.

Deformation with plane strain with prior uniaxial tension invites failure with lower major strains. These are higher than the limit strains in the constant strain ratio of FLD. Sonne et al. conducted experiments to study the strain path for aluminum-killed and titanium-stabilized steels. They showed that prestraining during drawing or uniaxial tension shifted the FLD in the direction of increasing major strain.

Predeformation with biaxial stretching resulted in a downward shift in the FLD. The researchers proposed a set of empirical rules for predicting forming limits with a two-stage strain path scenario, based on their experimental findings. Kleemola and Pelkkikangas also worked on the use of two-stage strain paths with either uniaxial or balanced biaxial prestraining. They also achieved similar results with experiments on AKDQ steel.

Second-stage strain ratios with uniaxial tension and plane strain result in the occurrence of failure at slightly lower major strain. Laukonis and Ghosh studied cold-rolled and annealed aluminum-killed steel and 2036-T4 aluminum alloy for equi-biaxially prestrained conditions. This resulted in a downward shift in the FLC for biaxially prestrained AK steel.

Lloyd and Sang experimented on forming limits of aluminum alloys, and studied strain path effects of five different aluminum alloys, including 1100-0, 3003-0, 2036-T4, 5182-0, and X-96. The tests were carried out in two steps: initially, the specimens were prestrained in uniaxial tension and then tensile deformation was applied in an orthogonal direction.

Many researchers attempted to develop forming limits based on strain paths. They applied the M–K theory for neck formation studies and the theory of plasticity for deformation. The first attempts to estimate failure curves for two-stage strain paths were made by Rasmussen. He experimentally validated the results on aluminum-killed steel. An upward shift was observed in the FLC for uniaxial prestraining but it was balanced for biaxial prestraining.

1.7 Construction of Forming-Limit Diagram

FLDs represent major ($\varepsilon 1$) and minor ($\varepsilon 2$) principal strains, as shown in Figure 1.6 on the x-axis and y-axis, respectively. The formability limit is usually characterized by the failure, that is, cracking, and is called the FLC.

But it is obvious that, in an actual production system, even necking cannot be allowed. It is important from the perspective of both functional aspects and aesthetics. So we must differentiate between the necking curve and the actual failure curve. The zone between the fracture and the necking-limit curve is called the range of local necking. Localized compressive stresses in the flange and wall cause accumulation of material, which is known as wrinkling.

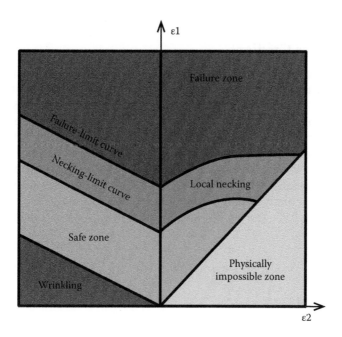

FIGURE 1.6
Failure-limit diagram—zones.

It is also evident that besides rupture and local necking, wrinkling must also be avoided. The FLD with these limit curves and zones is shown in Figure 1.6. The area below the necking, the green zone, indicates the safe region of normal forming conditions in terms of major ($\varepsilon 1$) and minor ($\varepsilon 2$) principle strains. Various limit curves and zones are represented by different colors to indicate the different behaviors from the point of view of formability. The fracture-limit curve, which is more often called the FLC, represents those limit values of $\varepsilon 1$ and $\varepsilon 2$ where fracture occurs. The necking-limit curve represents localized necking. The exact measurement of the principle strain components at necking and fracture is difficult. However, with recent developments in optical techniques, it is possible to directly generate an FLD by scanning the formed components.

The following are the most important parameters for experimentally plotting an FLD:

- The strain path, that is, the strain history applied during investigation
- The material and its properties
- Material thickness
- The shape and size of the specimen
- Experimental conditions, that is, the size of grid pattern, precision of grid measurement, and friction conditions

1.8 Thickness Gradient

With the application of a load, the metal flows and undergoes deformation. It perceives a necking region as soon as the load becomes equivalent to the tensile stress. The necking occurs due to extended thinning. This localized necking was observed to be independent of strain paths, speed of forming, and the material to be formed. The thinning and thickening refers to the behavior of sheet metal in positive and negative directions with respect to the original thickness at various cross sections. The thickness gradient refers to the ratio of neighboring nodes (or circles in practical formability analysis) in finite element analysis and this was experimentally estimated to be equal to or more than 0.92. The lesser-value elements (circles) are considered to be under necking.

Bibliography

Date, P. P., Sheet metal formability, *Proceedings of the Training Programme of Sheet Metal Forming Research Association*, Indian Institute of Technology, Mumbai, India, 2015.

Ghosh, A. K. and S. S. Hecker, Stretching limits in sheet metals: In-plane versus out-of-plane deformation, *Metallurgical Transactions* 5 (1974), 2161–2164.

Goodwin, G. M., Application of strain analysis to sheet metal forming problems in the press shop, Society of Automotive Engineers Technical Paper No. 680093, New York, 1968.

Hasford, W. H. and R. M. Cadell, *Metal Forming—Mechanics and Metallurgy*, Cambridge University Press, Cambridge, U.K., 2007.

Hecker, S. S., Simple technique for determining forming limit curves, *Sheet Metal Industries* 52 (1975), 671–676.

Hecker, S. S., Experimental studies of sheet stretch ability, *Formability: Analysis, Modeling and Experimentation: Proceedings of the Symposium at the AIME/ASM Fall Meeting*, Chicago, IL, October 1977, The Metallurgical Society of the American Institute of Mining, Metallurgical and Petroleum Engineers, 1978, pp. 150–182.

Hecker, S. S., A. K. Ghosh, and H. L. Gegel, Formability—Analysis modeling and experimentation, *Proceedings of a Symposium* held at Chicago, IL, sponsored by American Society of Metals Flow and Fracture, New York, 1977.

Hsu, T. C., W. R. Dowle, C. Y. Choi, and P. K. Lee, Strain histories and strain distributions in a cup drawing operation, *ASME Journal of Engineering for Industry* 93 (1971), 461–466.

Keeler, S. P., Determination of forming limits in automotive stampings, *Sheet Metal Industries* 42 (1965), 683–691.

Keeler, S. P., Circular grid system—A valuable aid for evaluating sheet metal formability, Society of Automotive Engineers Technical Paper No. 680092, New York, 1968.

Keeler, S. P. and W. A. Backofen, Plastic instability and fracture in sheets stretched over rigid punches, *ASM Transactions* 56 (1963), 25–48.

Kleemola, H. J. and M. T. Pelkkikangas, Effect of pre deformation and strain path on the forming limits of steel, copper and brass, *Sheet Metal Industries* 54 (1977), 591–599.

Kobayashi, T., H. Ishigaki, and T. Abe, Effect of strain ratios on the deforming limit of sheet steel and its application to actual press forming. *Proceedings of the Seventh Biennial Congress of the IDDRG*, International Deep Drawing Research Group, Amsterdam, the Netherlands, 1972, pp. 8.1–8.4.

Lang, K., *Hand Book of Metal Forming*, McGraw Hill Publications, New York, 1985.

Laukonis, J. V. and A. K. Ghosh, Effects of strain path changes on the formability of sheet metals, *Metallurgical Transactions* 9A (1978), 1849–1856.

Lee, D. and F. Zaverl, Jr., Neck growth and forming limits in sheet metals, *International Journal of Mechanical Sciences* 24(3) (1982), 157–173.

Lloyd, D. J. and H. Sang, The influence of strain path on subsequent mechanical properties—Orthogonal tensile paths, *Metallurgical Transactions* 10A (1979), 1767–1772.

Logan, R. W., Sheet metal forming—Simulation and experiments, PhD dissertation, University of Michigan, Ann Arbor, MI, 1985.

Marciniak, Z., J. L. Duncan, and S. J. Hu, *Mechanics of Sheet Metal Forming*, Butterworth Heinemann, Oxford, U.K., 1992.

Marciniak, Z. and K. Kuczynski, Limit strains in the process of stretch-forming sheet metal, *International Journal of Mechanical Sciences* 9 (1967), 609–620.

Marciniak, Z., K. Kuczynski, and T. Pokora, Influence of the plastic properties of a material on the forming limit diagram for sheet metal in tension, *International Journal of Mechanical Sciences* 15 (1973), 789–805.

Matsuoka, T. and C. Sudo, The effect of strain path on the fracture strain of steel sheets, *Sumitomo Search* 1 (1969), 71–80.

Memon, S., A. Omar, and K. Narasimhan, Finite element analysis for optimizing process parameters in tube hydro forming process, *IDDRG Conference*, Zurich, Switzerland, June 2–5, 2013.

Muschenborn, W. and H. Sonne, Influence of the strain path on the forming limits of sheet metal, *Archiv für das Eisenhuttenwesen* 46(9) (1975), 597–602.

Nikhare, C. P., Experiments and predictions of strain and stress based failure limits for advanced high strength steel, *Materials Today Proceedings* (2016), pp. 1–7.

Obermeyer, E. J. and S. A. Majlessi, A review of recent advances in the application of blank-holder force towards improving the forming limits of sheet metal parts, *Journal of Materials Processing Technology* 75 (1998), 222–234.

Pierce, R., *Sheet Metal Forming*, Adam Hilger, Bristol, U.K., 1991.

Rasmussen, S. N., Assessing the influence of strain path on sheet metal forming limits, *Proceedings of the 12th Biennial Congress of the IDDRG*, International Deep Drawing Research Group, Genoa, Italy, 1982, pp. 83–93.

Sang, H. and D. J. Lloyd, The influence of biaxial pre strain on the tensile properties of three aluminum alloys, *Metallurgical Transactions* 10A (1979), 1773–1776.

Tisza, M. and Z. P. Kovacs, New methods for predicting the formability of sheet metals, *Production Processes and Systems* 5(1) (2012), 45–54.

Vasin, R. A., F. U. Enikeev, and M. I. Mazurski, Method to determine the strain-rate sensitivity of a super plastic material from the initial slopes of its stress–strain curves, *Journal of Materials Science* 33(4) (February 1998), 1099–1103.

2

Process Parameters in Drawing

2.1 Blank-Holding Force

In the forming process, plastic deformation of the blank to the conformed shape is achieved by using a set of punch and die. The conversion of the blank into the finished product is dependent on various parameters. Defects are introduced into the component if the parameter selection is not proper. Therefore, optimum process parameters must obviously be selected to avoid incurring defects and to minimize costs and scraps. All parameters like machine parameters and process parameters are optimized with respect to their influence on the sheet metal forming characteristics.

The literature survey indicates that the blank-holding force plays a considerable role in the product quality and forming. An appropriate blank-holding force, selected through proper design, can result in controlling the thickness variations and strain distributions. An optimum blank-holding force controls wrinkling in the flange and walls and cracking at the bottom and wall region. Traditionally, a constant blank-holding force has been applied to minimize mechanisms in forming, but several researchers have experimented with varying values using hydraulic press. Let us take a detailed survey of the various contributions and developments in the strategy of applying a blank-holding force.

S. Aleksandrovic and M. Stefanovic revealed through experiments that there are advantages as well as disadvantages of applying a continuously varying blank-holding force. The blank-holding force acts as a normal force for friction and has significant influence on the process, primarily by preventing the appearance of wrinkles on the flange, and also by controlling the thinning in critical zones by balancing the radial and circumferential forces. The thinning process in the radial direction must be balanced with circumferential material accumulation. A significant advantage of the blank-holding force is the possibility of changing it in the course of process. The results obtained are indicative of the possibility of accomplishing an improvement in the quality of the process with a variable blank holder.

T. Yagami, K. Manabe, and Y. Yamauchi investigated the relation of the motion of a blank holder with the deep-draw ability of a circular cup of thin sheet metal by proposing a new algorithm. The perspective of investigation was to study wrinkling behavior as well as the fracture limit. A small blank-holding force is applied, with controlled motion, which allows temporary wrinkling, for improving the deep-draw ability of thin blanks. The algorithm was designed so that it achieved specific results. These authors also experimented on Cu alloy sheet to understand the influence of motion control on compressive stresses, resulting in wrinkles. Fracture damage reduction on the deep-drawing process was studied using finite element simulations. This also justified the use of the proposed methodology. The simulation work proved that this motion control strategy is best suited when experimenting with relatively thin sheet blanks. Wrinkles can be easily suppressed or eliminated if they are within the allowable limits of height. Therefore, minimum and maximum allowable wrinkle heights are control parameters that define wrinkle existence. The deep-draw ability is also enhanced as a result of reduction in ductile damage accumulation with this control.

R. Padmanabhan, M.C. Oliveira, J.L. Alves, and L.F. Menezes proposed a variable blank-holding force that would trace the lower boundary in the process in steps. In the first step, a small force is applied till formation of the wrinkles is observed. In the second step, the force is proportionally increased with the punch displacement. Proportionality depends on the material flow characteristics. The quality of the component was observed to improve significantly with this method.

H. Gharib, A.S. Wifi, M. Younan, and A. Nassef developed an optimization strategy that searches for blank-holding force, minimizing the maximum punch force and avoiding process limits. The strategy was applied to both constant and linearly varying blank-holding force. The conclusion was that the optimized linear blank-holding force scheme resulted in an improved cup forming. This was tried with diverse draw ratios and coefficients of friction at the die interface to understand and analyze the optimum linear blank-holding force. The slope of the linear blank-holding force scheme increases with the increase in the drawing ratio.

Kozo Osakada, Chan Chin Wang, and Ken-ichi Mori applied a new methodology for the full utilization of the latest presses that have the facility of a variable blank-holding force. Thus, an optimal history of the blank-holding force can be determined. The proposed finite element method (FEM) tries to find out the optimum blank-holding force, with which neither wrinkling nor localized thinning is observed during every deformation step. The applicability of the approach was demonstrated with stamping of axisymmetric deep-drawn parts of aluminum sheets, with obtained history.

Numerical simulations using Abacus Explicit to optimize different blank-holding forces at different locations of the blank-holder surface were conducted by Lanouar Ben Ayed, Arnaud Delamézière, and Jean-Louis Batoz. An optimization algorithm based on the response surface method was proposed, with minimization of the punch force as the ultimate objective. Three inequality constraints were defined for avoiding necking and wrinkling phenomena. This was applied to analyze front door panels. Better strain distributions were achieved in the drawn part. Punch displacement was controlled by using this methodology, resulting in the control of the blank-holding force.

A. Wifi and A. Mosallam conducted performance analysis of a few non-conventional blank-holding techniques applying the FEM. This included friction-actuated, pulsating, and pliable blank-holding techniques. They investigated the influence of various blank-holding force schemes on sheet metal formability limits, especially on wrinkling and tearing with explicit three-dimensional analysis. Three nonconventional blank holders were considered: friction actuated, elastic, and pulsating. Compared with the fixed blank-holding force scheme, it revealed that there is slight improvement in the formability with three blank-holder schemes under consideration.

The literature survey exhibits extensive research work for determining the optimum blank-holding force profile. Normally, deterministic approaches are applied that do not consider the inherent process variations, and lead to a highly unreliable design. So there is the need of a probabilistic approach, which was developed by Osueke and Ofondu, to take care of process variabilities. Gaussian functions were applied to the blank-holding force. Every parameter was varied with three levels (low, medium, and high) to test its sensitivity. All other parameters were kept at medium level. The variables were classified as control variables and noise variables to suit the design of the experiment. The Box Behkn DOE approach was applied for further improvement. With the default presence of the variations in sheet thickness and frictions between the sheet, die, binder, and punch, the probabilistic design successfully finds the optimal variable blank-holding force. Monte Carlo simulation with a sample size of 5000 was used for probability calculations. The yield saw improvement from 45.78% to 99.08%, revealing that the probabilistic design is robust as well as reliable.

An intelligent press control system with support of the FEM and database was applied to the sheet-stamping process of trapezoidal panels by Hiroshi Koyamaa, Robert H. Wagoner, and Ken-ichi Manabea. For non-symmetric shapes, a radial assignment strategy of the process control line was proposed. The entire process was controlled with the composition of each process control line. Virtual processing was implemented for better process control. A hydraulic forming simulator was used for experimentation along with virtual processing, for combined verification. High-quality products were manufactured with higher strength using a simulator with a variable blank-holding path having a virtual database and an FEM-availed perspicacious press control system. The control path was tenacious without assistance from an engineering expert, so efficiency of the process control was substantiated.

While executing deep drawing of rectangle components, in lieu of adopting elastic or several segmented blank holders, Zhu Wei, Z.L. Zhang, and X.H. Dong made an incipient suggestion of applying dual layers of blank holders that are parallel to each other to asynchronously vary blank-holding force at different regions of the sheet flange. This would help to supersede the little mandrills of different lengths between the dual layers of the blank holders, thus improving the formability of the products.

Concurrently, in order to solve the quandary of how to judge the desired distribution of the variable blank-holding force for asymmetric components, obtained in advance from the thickness analysis of the products, and to evaluate the best lubricant in deep drawing, a probe sensor with a structure of dual cantilevers, predicated on the theory of resistance strain slice, fine-tuned into the inner space of the die, was developed for quantifying the authentic-time friction coefficient of the sheet flange's surface. This special probe sensor made it possible to establish not only the real-time blank-holding force on the flange, but also the real-time friction coefficient.

Control of wrinkling during metal forming has been a major concern in recent years. In order to provide reliable and efficient tools that can predict the critical blank-holding force to obviate wrinkles, an axisymmetric analytical model for flange wrinkling was developed by H. Yoon, S. Alexandrov, K. Chung, R.E. Dick, and T.J. Kang. Using the conventional theory of critical condition, the critical blank-holding force is predicted numerically.

2.2 Friction in Drawing

The drawing operation consists of plastic deformation of sheet metal along with an intricate flow of material into the die, where friction plays a consequential role. The plane blank is composed and formed into the final product within the punch and die cavity. The sheet metal slides relative to the implement as well as to the blank holder, and friction occurs at every interaction. The sheet is stretched by the punch into the die, and by the friction between the punch and the sheet, to an astronomically immense extent, determining the deformation. Thinning transpires by this phenomenon alone. When the sheet slides over the curvature of the die, the sheet material is sheared. The friction between the die and the sheet influences the drawing force and thus the probability of composing without crack formation.

2.3 Frictional Problems in Deep Drawing

The level of friction that occurs in a drawing operation is a crucial parameter. The free flow of material resulting in wrinkling due to low friction and high friction lead to tearing and crack formation. The sheet material adheres to the tool surface particularly in draw beads and sharp radii, so it is very difficult to obtain the exact level of friction. This phenomenon, often called galling, causes a variable friction from one part to the next and interrupts production as the tool surfaces need cleaning.

2.4 Contact Regions in Deep Drawing

When two bodies are in contact and a relative motion subsists between these bodies, friction arises. In deep drawing, friction originates from contact between the implement and the sheet. Schey distinguishes a total of six contact and friction regions in deep drawing. These regions are shown in Figure 2.1. Regions 1 and 2 are the contact regions between the blank holder and the sheet, and the die and the sheet, respectively. These two regions together are called the flange. In the flange region, radial drawing between the die and the blank holder occurs and the strains in the sheet are rather minute. The nominal pressure in the flange region is low, that is, on the order of 1–10 MPa. Region 3 represents the contact between the die rounding and the sheet. In this region, the sheet is bent and unbent. High nominal pressures on the order of 100 MPa occur in this region. The tension force is high and stretching plays a consequential role. The contact between the punch and the

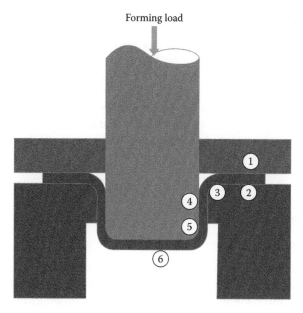

FIGURE 2.1
Contact regions in deep drawing.

sheet is found at region 4. The sheet is stretched further in this region but no authentic physical contact occurs. In contact region 5, contact between the punch radius and the sheet occurs. In region 6, contact between the bottom of the punch and the sheet occurs, causing the sheet to stretch. In deference to friction, the literature does not report much about the regions of contact between the punch and the sheet and the punch bottom and the sheet, because the friction in these regions does not influence the deep-drawing process much. The friction in the other regions, that is, the flange region, the die radius region, and the punch radius region, influences the deep-drawing mechanism in different ways. The friction in region 5 must be amply high to ascertain that the sheet follows the kineticism of the punch. The friction in regions 1, 2, and 3 must not be too high because a high friction leads to higher punch forces, resulting more facilely in fracture.

2.5 Lubrication Mechanism in Deep Drawing

The contact regions 1, 2, and 3, mentioned in Section 2.4, are important in deep-drawing processes, and operate in different lubrication regimes. In the flange region, the pressure is low (below 10 MPa) because of the large contact area. This implies that hydrodynamic effects may play a role in lubricating the contact and that lubrication takes place in the low mixed-lubrication (ML) region. However, as observed by Emmens, during deep drawing, the pressure in the flange region increases due to a decreasing contact area. Thickening of the outer part of the blank causes a more concentrated contact in this region. In the contact region between the die radius and the sheet, the pressure is much higher (around 100 MPa) than in the flange region. Therefore, the conditions are more severe in this region. The same situation arises for the contact between the punch radius and the sheet. The pressure is high, which implies the occurrence of boundary lubrication and ploughing (BL&P). It can be concluded that BL&P are important lubrication mechanisms in deep drawing.

2.6 Micromechanism

The predominant lubrication mechanism and the mechanisms controlling the micro contacts between the sheet and the tool decide the friction level. When the tool and the sheet move relative to each other, grinding scratches are pressed into the sheet surface to form a plastic wave. Alternatively, and depending on the geometry of the surface roughness, cutting and/or scratching may occur. Independently of which mechanism predominates, sliding is prevented by this deformation mechanism. Plastic waves are demonstrated during sliding in the vertical walls. One of their components might have transformed into cutting. The influence of the size and shape of the contact areas on the friction level is demonstrated by the fact that the micro mechanisms that brake the sliding of a sheet all contribute to this effect at each point of contact. Ergo, the total friction force depends on the total contact area and this, in turn, depends on the surface roughness of the sheet. However, the analysis is complicated by the fact that both the pristine surface roughness and the surface transmuted after deformation need to be compared. Change in surface roughness due to deformation is known as the orange peel effect. Cognizance of these mechanisms brings the entelechy that the mechanical characteristics of a sheet surface are of great consequence for the friction level, both due to the braking force of plastic waves and because of the promotion of the contact areas by plastic deformation of the surface. Deciding which lubricating mechanism is to predominate is a matter of how thick a lubricating film can be maintained in cognation to the surface roughness of an implement and sheet. If the lubricating film is adequately thick to obviate contact between the sheet and the implement surfaces, "full film" lubrication can be applied, but this is infrequently perceived in sheet composing. Consequently, mixed lubrication and "boundary" lubrication are most mundane. In the first case, there is an adequate amount of lubricant to fill the depressions in the sheet surface so that the flow of lubricant to the contact areas can be availed to build up protective layers, thus reducing friction in decrements. The lubricant can be applied with pressure, so a component of the contact pressure is carried by the lubricant in the depressions. Only a modicum of lubricant is applied on the surface in boundary lubrication. The contact pressure is carried by the points of contacts.

2.7 Generalized Stribeck Curve

For the case of two lubricated surfaces which slide along each other under a normal load, three different lubrication regimes can be distinguished:

1. *Boundary lubrication (BL)*: The normal load is carried totally by the contacting asperities, which exist on both surfaces. These surfaces are protected from dry contact by thin boundary layers, which are attached to the surfaces.

2. *Mixed lubrication (ML)*: A part of the load is carried by contacting asperities (separated by boundary layers) and another part of the load is carried by the lubricant film.

3. *Hydrodynamic lubrication (HL)*: The load is carried totally by the full film and contact between the opposing surfaces does not occur. When the normal load is high, elastic deformation of the surfaces may occur. In this case, the term elasto-hydrodynamic lubrication is used to define the lubrication mechanism.

2.8 How to Control Friction Level

The control rather than minimizing of friction gives better output in sheet metal forming. It is usually a good method to use small amounts of lubricant, typically 1 g/m², to achieve a friction level that is relatively insensitive to variations in the pressing speed. So boundary lubrication is preferred over others. For certain combinations of materials, there is a great risk of adherence to tool surfaces, resulting in high friction levels. So, when applying this method, tool geometry must be accommodated for high friction levels in order to have a stabilized state. The combination of sheet material, tool material, and lubricant should counteract adherence. For achieving a low friction level, more lubricant must be applied of higher viscosity; this brings about the concept of mixed lubrication. The concern with low-level friction is that it becomes sensitive to variations in the forming process, that is, ram speed, amount of lubricant, blank-holding force, and so on.

2.9 Ploughing and Adhesion

A simple Coulomb friction model is still popular with researchers. The influence of various parameters on contact behavior is not accounted for in Coulomb's model. These parameters include pressure, speed of punch, as well deformation. Consequently, to predict the draw-in capability of sheet metal and springback with the latest models is cumbersome. Several experimental and theoretical attempts have been made to understand contact and friction information of lubricated sheet metal. At the microscopic level, friction is due to adhesion between contacting asperities. Ploughing of asperities and adhesion effects within boundary layers mainly cause friction in the boundary region. If the contact pressure is carried by the asperities and lubricant flow, as in the ML regime, or if it is fully carried by the lubricant, as in the HL regime, hydrodynamic shear stresses will contribute or even predominate. The following discussion will focus on the two friction mechanisms present in the boundary layer regime: ploughing and adhesion. Wilson developed a model that treated the effect of adhesion and ploughing separately. A more advanced model was developed by Challen and Oxley, which takes the combined effect of ploughing and adhesion on the coefficient of friction into account. Challen and Oxley experimented with a soft flat material and applied hard wedge-shaped asperity, and expressions were developed for coefficient of friction and wear rates. Westeneng extended Challen and Oxley's model to describe friction conditions between multiple tool asperities and a flat work piece material. Their model considers the flattened plateaus of the work piece asperities as soft and perfectly flat, and the surface texture of the tool as hard and rough.

The real area of contact determines the proportion of ploughing and adhesion. So, the coefficient of friction is dependent on the real area of contact. Various flattening and roughening mechanisms define the real area of contact. Flattening due to normal loading, flattening due to stretching, and flattening due to sliding are the predominant flattening phenomena. It increases the real area of contact, resulting in a higher coefficient of friction. The asperities are roughened during stretching of the deformed material, which results in a lower coefficient of friction.

Many researchers have contributed to friction modeling, especially in the development of models, with the objective of predicting the flattening behavior of asperities due to normal loading. The majority were inspired by the pioneering work of Greenwood and Williamson. They proposed a stochastic model based on the contact between a flat tool and a rough work piece surface. The model is a mathematical description of the contact between two surfaces. It is based on the assumption that particles of the rough surface are spherical, they only deform elastically, and the surface texture can be described by a distribution function.

2.10 Friction and Drawability

The term drawability is complex enough to define precisely. It is dependent on material properties and interaction of tool and metal sheet. Drawability cannot be predicted based only on properties measured by standard tensile tests. Boundary conditions like friction distribute the strain throughout the sheet surface. Limit drawing ratio (LDR) and limit strains are two indicators of drawability of a material. Different strain distributions are observed in different regions of a sheet, due to varying lubrication. It has also been pointed out that inaccuracies of friction values sometimes have more profound effects on simulation results than most material properties. Paunoiu et al., while plotting the forming-limit curves for a steel specimen, observed that the limit strains with deep-drawing oil as lubricant differed largely from those when thin film of polymer and oil were used. This strengthens the argument that the limit strains depend upon the friction coefficient.

The coefficient of friction is usually determined by nominal pressure, bulk strain, and hardness. The distribution of asperity height on the work piece surface and the roughness parameters of the tool and the boundary lubricant also matter. Since optimum blank-holding force and punch load are interrelated, the normal pressure varies with the punch travel. Moreover, bulk stretching also influences the topography of the work piece. When simultaneous normal loading and stretching are applied, the work piece surface may be fine-tuned or roughened, depending on the thickness of the material and the applied nominal pressure, thereby varying the coefficient of friction. But boundary layer lubrication in combination with ploughing of tool asperities with the work piece is an important friction mechanism in deep-drawing processes, where the coefficient of friction is assumed to be constant. Many researchers have contributed with their insights into the field of friction.

G. Chandra Mohan Reddy et al. studied the effect of the coefficient of friction on drawability using finite element analysis. The model was validated with the comparison of forces during experimentation and simulation. The study reveals that the LDR decreases linearly with the increase of coefficient of friction.

Wilko C. Emmens studied the frictional aspects of aluminum and concluded through the results of his experiments that aluminum shows a strong flattening of roughness asperities by a sliding contact. With ML, the flattening of asperities has an impact on pressure.

In the case of aluminum, we cannot shift from ML to BL just by increasing the pressure, as the product often does not fracture even at the highest blank-holding force.

Friction at the tooling–work piece interface is very important in sheet metal–forming operations. Accurate knowledge of friction at such interfaces is needed for process design and analysis, numerical process simulation, and validation to control forming processes.

S. Hao, B.E. Klamecki, and S. Ramalingam developed two physical models or friction simulators based on the stretching of a strip around a pin, to characterize sheet metal–forming friction. Direct measurements of the forces were proposed, instead of strain, to infer friction forces.

Luo Yajun and H.E. Dannong researched friction law in the rectangular parts of deep drawing using a simulation test machine that assesses lubricant performance in deep-drawing processes. Friction coefficients at different positions on the die were measured with probe sensors. Friction coefficients were described and quantified at the flange corner, near the straight border, and at the far straight border. This facilitated the selection of appropriate lubricants in auto-body panel drawing.

O. Mahrenholtz, N. Bontcheva, and R. Iankov studied the influence of roughness at contact surfaces by means of a nonlocal friction law. The effect of the density of asperity distribution on normal compliance law was studied, with different 3D models. The effect of normal, as well as tangential, tool velocity was investigated. It was observed that the tangential tool velocity had no influence on the normal compliance law. These authors proposed a model for predicting the dependence of the Coulomb friction coefficient on surface roughness.

2.11 Process Improvements

From previous discussions, it is clear that forming is a complex manufacturing process involving a large number of parameters. Their interactions determine the success or failure of the product. Traditionally, it was considered an art rather than a science, as it depends on the skill, knowledge, and intuition of the designer or operator. The experimentation is lengthy and involves high costs, long time, and considerable human efforts. Many researchers have attempted to study the process as a system and have tried to optimize the process in different forms. A few of these attempts are presented here.

Juraj Hudak and Miroslav Tomas contributed in the research of force parameters, by experimenting with drawing and blank-holding force in the deep-drawing process of a flat-bottomed cylindrical cup. They used steel sheets for enameling KOSMALT produced by U.S. Steel Kosice, Ltd. The process was complicated due to the contradiction between the steel structure's good drawability and good enameling. The measurement of the force parameters (drawing and blank holding) was realized through an elastic deformation element called the dynamometer, which is equipped with four tensometric sensors connected to Wheatston's bridge.

LDRs and other axisymmetric deep-drawing characteristics of AA5754-O and AA6111-T4 automotive aluminum sheet materials were investigated as a function of die profile radii with experiments and numerical predictions by M. Jain, J. Allin, and M.J. Bull. They proposed an approach for rapid determination of LDR depending on the characteristic limit load of the material at the fracture. It was observed that the die profile radius has an impact on LDR and flange draw-in by reducing them. This decrease is attributed to the increased work hardening in the die profile region. Additional bending, unbending, and stretching of the material entering the die cavity, as well as the increased tendency toward the fracture in AA6111-T4, causes this.

Sergey F. Golovashchenkoa, Nicholas M. Bessonov, and Andrey M. Ilinich proposed a two-step method to form a part and design a preform shape. The part may be formed from lightweight material to an extent that would normally exceed the forming limits of the material if the part were attempted to be formed in a one-step conventional stamping die. Critical areas including deep pockets and sharp radii of the final part can be formed from a preform or intermediate shape part. The first step can be manufactured by a variety of sheet metal–forming methods; the preformed blank is further formed in a fluid pressure process to arrive at the final part shape, wherein broad radius areas and pockets of accumulated metal of the preform are formed into deep pockets and sharp corners of the final shape. Electro-hydraulic forming technology is employed for the second forming step.

I. Karabegovic and E. Husak proposed a mathematical model for deep-drawing force with double reduction of wall thickness. Deformation, diameter ratio, and friction coefficient were studied experimentally with Taguchi design of experiments. This experimentation was done using a hydraulic testing machine, Amsler 300 KN. The raw specimen is CuZn28, which was prepared before the deep-drawing process by heating, washing, and lubricating. The results obtained indicate that the final mathematical model is adequate for describing the process of deep drawing, which is confirmed by Fisher's test. The goal of optimization with the mathematical model is to get minimal deep-drawing force with the reduction of wall thickness with optimal input parameters (factors). Force minimizing has multiple facets: from minimizing energy consumption to decreasing intensity of wasting on press die guides and other pieces of the press die.

Nedeljko Vukojević et al. present the optimization of geometric parameters in deep-drawing tools to reduce stress concentrations. Optimization is performed through numerical analysis using parametric optimization.

Optimization of process parameters in hydro-mechanical deep drawing (HDD) of a trapezoid cup is an important task for reducing production cost. To determine the optimum values of the process parameters, it is necessary to determine their influence on the deformation behavior of the sheet metal. Thickness distribution of the trapezoid cup depends on punch speed, chamber pressure, and friction coefficient.

Thanh-Phong Dao and Shyh-Chour Huang proposed a combination of FEM and Taguchi method to predict the influencing parameters. The influence of each parameter was studied. Numerical simulation of orthogonal array (L9) was carried out. It was found that friction efficiency has the greatest effect on thickness distribution in the HDD of the trapezoid cup.

Prediction of successful forming is important. M.M. Moshksar and A. Zamanian conducted a series of cup-drawing tests to study the deep drawing of commercial aluminum blanks. The critical die and punch shoulder radii, the limiting blank diameters, and the limiting drawing ratios were measured. Over the ranges of conditions investigated, the drawing process was found to be strongly sensitive to the die and punch-nose radii.

An important problem in the production of drawn parts is tool wear, especially at the draw-die. If tool wear can be reduced, this can increase the tool life as well as facilitate continuous production flow, by virtue of reduction in the number of breakdowns during the repolishing of tools.

M.R. Jensen, F.F. Damborg, K.B. Nielsen, and J. Danckert attempted to reduce tool wear using FEM and optimization technique to redesign the geometry of the draw-die profile. The optimized draw-die profile has almost twice the tool life compared to that of the initial circular draw-die if the peak value of wear is used as the wear criteria. With an elliptic draw-die profile, an improved shape of the cup wall is achieved.

H. Naceur et al. developed a one-step FEM called the inverse approach (IA) to estimate the large elastoplastic strains in thin sheet metallic parts obtained by deep drawing. In this work, the development of the IA for the analysis and optimum process design of industrial parts is presented. The first part of the research deals with improvements in the analytical formula to consider the restraining forces due to the draw beads. The second part deals with the optimization procedure combining the IA, a BFGS algorithm, and analytical sensitivity analysis to optimize material parameters and restraining forces.

One of the most important, yet simplest, deep-drawing operations on a flat blank of sheet metal is the production of a cylindrical cup. To further investigate this process, an instrumented tool was designed by M.T. Browne et al. for operation on a Norton, 20 ton, double-acting hydraulic press. The research work investigated the variation and effects of punch and die geometry, blank-holding pressure, top-ram pressure, lubrication, and drawing speed in the deep drawing of C.R.1 steel cups of 0.9 mm thickness. These effects were investigated by drawing a series of cups using DOE, where a screening experiment was conducted and the desired factors were varied at different levels. The chosen responses were punch load and wall thickness variation, which was conducted in order to investigate these effects and provide optimum levels for each of the factors.

H. Naceur et al. presented a new numerical approach to optimize the shape of the initial blank, which is formed into a final 3D work piece by the deep drawing of thin sheets. This proposed approach has two steps: first, IA is used for the forming simulation and, second, the evolutionary algorithm is applied. The proposed approach has shown potential enhancements in optimization of the blank contour for square cups.

Tsutao Katayama et al. focused on forming defects (fracture and wrinkle) in the two-stage deep drawing. To improve the quality of the process, fracture and wrinkle must be simultaneously controlled. For this reason, they proposed a transfer forming technique, which includes a new intermediate-process die shape. They investigated the influences of forming defects on the intermediate-process die shape, and searched an optimum die shape for improving both forming defects by using the multi-objective function and sweeping simplex method, which uses optimization. The research concluded with a new method to change the intermediate die shape into the two-stage deep drawing.

The work carried out by G. Gantar explores the possibilities of evaluating and increasing the stability of the deep-drawing process. The stability of production processes is verified with numerical simulation. This presents a new approach for optimization of the deep-drawing processes. The approach is capable of determining the optimal values of input parameters for the highest stability of the process. This was successfully applied to deep drawing of rectangular boxes.

Molybdenum remains mechanically strong at low as well as high temperatures. It is a widely applied refractory material with high heat and electrical conductivity. It has plenty of scope, especially in high-temperature applications. The only disadvantage is that it has very low drawability, so a multistage process must be designed. Heung-Kyu Kim studied a multistage deep-drawing process on a circular cup for a molybdenum sheet by including ironing, which was effective in increasing drawability. The effects of die design variables were studied with parametric finite element study. The design variables of the multistage deep-drawing process were selected after parametric study. A global as well as a local optimum search algorithm was applied to optimize the nonlinear process.

The optimization method and numerical simulation technology have been applied in the sheet metal–forming process to improve the design quality and shorten the design cycle. However, deterministic optimization may lead to unreliable and nonrobust designs due to overlooking the fluctuation of design variables, environments, and operational conditions. In addition, iterations in the optimization process may require considerable simulation time or great experiment cost.

A combined computer-aided engineering and six sigma robust design approach has been proposed by D.J. Zhang with the objective of eliminating uncertainties in design quality improvement. By applying the design of experiment, analysis of variance, and dual response surface model, this integrates the design for six sigma (DFSS), reliability optimization, and robust design together for optimization. Deep-drawing processes of a square cup (NUMISHEET 93) and a cylindrical cup (NUMISHEET 2002) are experimented with an objective function using thickness variation and wrinkle and rupture criteria as constraints. The results indicate significant improvement in the reliability and robustness of the forming quality as well as an increase in design efficiency with an approximate model.

The effect of using a tractrix-shaped die profile instead of the normal circular die profile in the deep drawing of cylindrical cups has been investigated experimentally and theoretically (using the explicit FEM-code Dyna2d) by Joachim Danckert. The experiments and FEM simulations show that by using a tractrix-shaped die profile in a one-stage deep-drawing process, a substantial increase in the straightness of the cup wall and a substantial decrease in the residual stresses can be achieved. With a tractrix-shaped die profile, the outside diameter of the cup is larger than the inside diameter of the draw-die, whereas the opposite is true when using a circular die profile. Experiments and FEM simulations show that by using a tractrix-shaped draw-die in a two-stage deep-drawing process, a substantial increase in the straightness of the cup wall can be achieved. When using a tractrix-shaped die profile in the second stage, the outside diameter of the cup is larger than the inside diameter of the draw-die; besides, from the very rim of the cup, the opposite is true when using a circular die profile in the second stage.

Based on a one-step simulation algorithm and the program developed, a new FEM was introduced by Xiaoxiang Shi et al. for the direct prediction of blank shapes and strain distributions for desired final shapes in sheet metal forming. Since a one-step approach needs to predict the initial blank shape and the thickness distribution in a deformed part rapidly, the application of the DKT shell element including bending effects is preferred because of its simplicity and effectiveness.

2.12 Die Profile Radius and Punch-Nose Radius

The die profile radius and the die-face surface are probably the most influential features in a draw tool that uses a flat blank holder. If the draw radius is too small, the part may split as the material deforms. This is due to the high restraining forces caused by bending and unbending of the sheet metal over a tight radius. Drawing over a tight radius also produces a tremendous amount of heat. This can lead to microscopic welding of the sheet metal to the tools, known as galling. On the other hand, an excessive die radius causes the blank to wrinkle in the unsupported region between the punch face and the die face. When the blank wrinkles, the engineered clearance between the punch and die cavity is reduced by the wrinkle height, material flow is impeded, and fractures in the stamping result. It is apparent that there must be some range of die radii to select from that will work which are not too small or too big.

The draw punch applies the required force on the sheet metal blank in order to cause the material to flow into the die cavity. The critical features of the draw punch include the punch face and the punch-nose radius. The punch-nose radius cannot be too small as it will try to pierce or cut the blank rather than force the material to bend around the radius. The minimum punch-nose radius depends on material type and thickness. It is equally important to understand that as the punch-nose radius is increased, the blank will tend to stretch on the punch face rather than draw in the blank edge. A large radius, especially one that is highly polished, reduces the amount of friction on the punch-face surface. Reducing friction here allows the material to stretch more easily across the punch, the path of least resistance, instead of drawing in the blank edge. When a large punch radius is required, it is often helpful to leave the punch face rough. This increases the coefficient of friction over the punch-face surface and discourages material flow, thus helping to pull in the blank flange toward the die cavity.

Recently, the numerical simulation of sheet metal–forming processes has become a very powerful tool in the automotive industry. It is applied to check part geometry at an early stage of the design, as well as to optimize the shape parameters for safe production. Xiaoxiang Shi et al. proposed a comprehensive approach to effect die shape optimization for sheet metal–forming processes. The objective is to optimize stamping quality by minimizing the risk of rupture, wrinkles, and unstretched areas. The design variables include shape perturbation vectors on addendum surfaces and draw bead height on binder surfaces. A modified sweeping simplex algorithm is adopted in the optimization process. The forming process of a front fender is used to validate this proposed approach.

If the final part is to be produced without any defects, the development process has to be supported by means of numerical simulations based on FEM. The experiences gained with optimization of sheet metal–forming processes are presented by Gasper Gantar et al. on industrial examples. The examples are carefully selected in order to present all important issues concerning sheet metal forming: determination of optimal product shape and optimal initial blank geometry, prediction of fracture, prediction of final sheet thickness, prediction of wrinkling, prediction of loads acting on the active tool surfaces, prediction of springback, and residual stresses in the product. The results of the numerical simulations were compared to the samples from the production processes.

In order to improve the fact that a satisfying result is often obtained by means of CAE software's repeated calculation and the designer's multiple adjustments, and to gain the process parameters in the least time, an automated optimization integration system of the sheet metal–forming process parameters was built based on the multidisciplinary design optimization platform, based on an insight by Qiang Liu. The system has a seamless interaction between the CAD and CAE software tools, which can improve the automated degree of the design and analysis, and greatly reduce people's intervention and repeated work. The building method of the optimization system is simple and practical. Finally, two samples are applied to verify the feasibility and validity of the proposed integration system.

T. Jansson and L. Nilsson proposed the use of optimization to evaluate alternative sheet metal–forming processes. Six process setups were first defined in a hierarchy of designs and optimization was then used to evaluate each forming process of these designs. The challenge in designing the forming process was to avoid failure in the material and at the same time reach an acceptable strain through the control of thickness. The conclusion of this study is that there may exist a different process that can give an improved product for the desired geometry. It might be impossible for the optimization algorithm to arrive at this process due to a poor starting point or a not-so-wise process setup.

Bibliography

Aleksandrovic, S. and M. Stefanovic, Significance and limitations of variable blank holding force application in deep drawing process, *Tribology in Industry* 27(1 and 2) (2005), 48–54.

Ayed, L. B., A. Delamézière, J.-L. Batoz, and C. Knopf-Lenoir, Optimization of the blank holding force with application to Numisheet'99 front door panel, *VIII International Conference on Computational Plasticity, COMPLAS VIII*, Barcelona, Spain, 2005.

Browne, M. T. and M. T. Hillery, Optimizing the variables when deep-drawing C.R.1 cups, *Journal of Materials Processing Technology* 136 (2003), 64–71.

Challen, J. and P. Oxley, An explanation of the different regimes of friction and wear using asperity deformation models, *Wear* 53 (1979), 229–243.

Chandra Mohan Reddy, G., P. V. R. Ravindra Reddy, and T. A. Janardhan Reddy, Finite element analysis of the effect of coefficient of friction on the drawability, *Tribology International* 43 (2010), 1132–1137.

Danckert, J., The influence from the profile in the deep drawing of cylindrical cups, Department of Production, Aalborg University, Aalborg, Denmark, 1996.

Dao, T.-P. and S.-C. Huang, Study on optimization of process parameters for hydro mechanical deep drawing of trapezoid cup, *Journal of Engineering Technology and Education* 8(1) (March 2011), 53–71.

Emmens, W. C., Some frictional aspects of aluminum in deep drawing, *International Congress on Tribology of Manufacturing Processes*, Gifu, Japan, pp. 114–121, 1997a.

Emmens, W. C., Tribology of flat contacts and its application in deep drawing, PhD thesis, University of Twente, Enschede, the Netherlands, 1997b.

Gantar, G., K. Kuzmana, and B. Filipi, Increasing the stability of the deep drawing process by simulation-based optimization, *Journal of Materials Processing Technology* 164–165 (2005), 1343–1350.

Gharib, H., A. S. Wifi, M. Younan, and A. Nassef, Optimization of the blank holding force in cup drawing, *Journal of Achievements in Materials and Manufacturing Engineering* 18(1–2) (September–October 2006), 291–294.

Golovashchenkoa, S. F., N. M. Bessonov, and A. M. Ilinich, Two-step method of forming complex shapes from sheet metal, *Journal of Materials Processing Technology* 211 (2011), 875–885.

Greenwood, J. and J. Williamson, Contact of nominally flat surfaces, *Proceedings of the Royal Society of London. Series A, Mathematical and Physical Sciences* 295 (1966), 300–319.

Hao, S., B. E. Klamecki, and S. Ramalingam, Friction measurement apparatus for sheet metal forming, *Wear* 224 (1999), 1–7.

Hola, J., M. V. Cid Alfarob, M. B. de Rooijc, and T. Meindersd, Advanced friction modeling for sheet metal forming *Wear* 286–287 (May 2012), 66–78.

Hudak, J. and M. Tomas, Analysis of forces in deep drawing process, *Annals of Faculty Engineering Hunedoara—International Journal of Engineering* 9(1) (2011).

Jain, M., J. Allin, and M. J. Bull, Deep drawing characteristics of automotive aluminum alloys, *Materials Science and Engineering* A256 (1998), 69–82.

Jensen, M. R., F. F. Damborg, K. B. Nielsen, and J. Danckert, Optimization of the draw-die design in conventional deep-drawing in order to minimize tool wear, *Journal of Materials Processing Technology* 83 (1998), 106–114.

Karabegovic, I. and E. Husak, Mathematical modeling of deep drawing force with double reduction of wall thickness, *Mechanika* 2 (70) (2008), 61–66.

Katayama, T., E. Nakamachi, Y. Nakamurac, T. Ohata, Y. Morishita, and H. Murase, Development of process design system for press forming—Multi-objective optimization of intermediate die shape in transfer forming, *Journal of Materials Processing Technology* 155–156 (2004), 1564–1570.

Kim, H.-K. and S. K. Hong, FEM-based optimum design of multi-stage deep drawing process of molybdenum sheet, *Journal of Materials Processing Technology* 184 (2007), 354–362.

Koyamaa, H., R. H. Wagoner, and K.-i. Manabea, Blank holding force control in panel stamping process using a database and FEM-assisted intelligent press control system, *Journal of Materials Processing Technology* 152 (2004), 190–196.

Li, Y. Q., Z. S. Cui, X. Y. Ruan, and D. J. Zhang, CAE-based six sigma robust optimization for deep-drawing sheet metal process, *International Journal of Advanced Manufacturing Technology* 30 (2006), 631–637.

Mahrenholtz, O., N. Bontcheva, and R. Iankov, Influence of surface roughness on friction during metal forming processes, *Journal of Materials Processing Technology* 159 (2005), 9–16.

Moshksar, M. M. and A. Zamanian, Optimization of the tool geometry in the deep drawing of aluminum, *Journal of Materials Processing Technology* 72 (1997), 363–370.

Naceur, H., A. Delaméziere, J. L. Batoz, Y. Q. Guob, and C. Knopf-Lenoir, Some improvements on the optimum process design in deep drawing using the inverse approach, *Journal of Materials Processing Technology* 146 (2004a), 250–262.

Naceur, H., Y. Q. Guob, and J. L. Batoz, Blank optimization in sheet metal forming using an evolutionary algorithm, *Journal of Materials Processing Technology* 151 (2004b), 183–191.

Osakada, K., C. C. Wang, and K.-i. Mori, Controlled FEM simulation for determining history of blank holding force in deep drawing, *Annals of the CIRP* 44(1) (1995), 243–246.

Osueke, C. O. and I. O. Ofondu, Design for uncertainties in deep drawing processes using the probabilistic approach with an optimum blank holding force (BHF) profile, *International Journal of Advanced Engineering Sciences and Technologies* 8(2), 145–151.

Padmanabhan, R., M. C. Oliveira, J. L. Alves, and L. F. Menezes, An optimization strategy for the blank holding force in deep drawing, *Eighth World Congress on Computational Mechanics (WCCM8) and the Fifth European Congress on Computational Methods in Applied Sciences and Engineering (ECCOMAS 2008)*, Venice, Italy, June 30–July 5, 2008.

Paunoiu, V., D. Nicoar, A. Maria Cantera Lopez, and P. A. Higuera, Experimental researches regarding the forming limit curves using reduced scale samples, The Annals of "Dunarea De Jos" University of Galati Fascicle V, Technologies in Machine Building, ISSN 1221-4566, 2005, pp. 60–64.

Schedin, E., Control of friction in sheet metal forming can result in more stable production, *Materials & Design* 14(2) (1993), 127–129.

Schey, J. A., *Tribology in Metalworking—Friction, Lubrication and Wear*, ASM, Metals Park, OH, 1984.

Shi, X., Y. Wei, and X. Ruan, Simulation of sheet metal forming by a one-step approach: Choice of element, *Journal of Materials Processing Technology* 108 (2001), 300–306.

Vukojević, N., M. Imamović, and T. Muamer, Geometry optimization of tools for deep drawing, *14th International Research/Expert Conference "Trends in the Development of Machinery and Associated Technology" TMT 2010, Mediterranean Cruise*, Venice, Italy, September 11–18, 2010.

Wagoner, R., Development of OSU formability test and OSU friction test, *Journal of Material Processing Technology* 45 (1994), 13–18.

Wei, Z., Z. L. Zhang, and X. H. Dong, Deep drawing of rectangle parts using variable blank holding force, *International Journal of Advanced Manufacturing Technology* 29 (2006), 885–889.

Westeneng, J. D., Modeling of contact and friction in deep drawing processes, PhD thesis, University of Twente, Enschede, the Netherlands, 2001.

Wifi, A. and A. Mosallam, Some aspects of blank-holding force schemes in deep drawing process, *Journal of Achievements in Materials and Manufacturing Engineering* 24(1) (September 2007), 315–323.

Wilson, W., Friction models for metal forming in the boundary lubrication regime, *American Society of Mechanical Engineers* 10 (1988), 13–23.

Yagami, T., K. Manabe, and Y. Yamauchi, Effect of alternating blank holder motion of drawing and wrinkle elimination on deep-drawability, *Journal of Materials Processing Technology* 187–188 (2007), 187–191.

Yajun, L., H. E. Dannong, Y. Jianhua, Z. Yongqing, Z. Wei, and L. Jun, Research on friction law in deep drawing process of rectangular parts, *Science in China* 44(Series A), (August 2001), 230–234.

Yoon, H., S. Alexandrov, K. Chung, R. E. Dick, and T. J. Kang, Prediction of critical blank holding force criterion to prevent wrinkles in axi-symmetric cup drawing, *Material Science Forum* 505–507 (2006), 1273–1278.

3

Failures in Drawing

3.1 Thinning and Thickening Behavior

Thinning refers to reduction in thickness from the original thickness of the blank at various cross sections of the components under study. Thinning is measured from the simulation results of thickness at various cross sections, where it has reduced from the original, and the average reduction in thickness is calculated. Average reduction in thickness is used for analysis of variance.

Thickening is the increase in thickness at certain locations in the components under study due to compressive stresses in circumferential direction. Thickening is measured from the simulation results of thickness at various cross sections, where it has increased, and average thickening is calculated. The average increase in thickness is used for analysis of variance and further analysis.

3.2 Wrinkling in Forming

Wrinkling is usually undesired in final sheet metal parts for aesthetic or functional reasons. It is unacceptable in the outer skin panels where appearance of the final part is crucial. Wrinkling on the mating surfaces can adversely affect the part assembly and part functions, such as sealing and welding. In addition, severe wrinkles may damage or even destroy dies. There are three types of wrinkles that frequently occur in the sheet metal–forming process: flange wrinkling, wall wrinkling, and elastic buckling of the undeformed area owing to residual elastic compressive stresses. In the forming operation of stamping a complex shape, draw-wall wrinkling means the occurrence of wrinkles in the die cavity. Since the sheet metal in the wall area is relatively unsupported by the tool, the elimination of wall wrinkles is more difficult than the suppression of flange wrinkles. For the same reason, the magnitude of the compressive stress necessary to initiate side-wall wrinkling is usually smaller than that for flange wrinkling. Hence, the formation of side-wall wrinkles is relatively easier, especially when the ratio of the unsupported dimension to sheet thickness is large.

In addition, the trim line of the part is usually located a little inside the die radius, and only the wrinkling in the frustum appears in the final part. Hence, side-wall wrinkling is a problem of greater industrial importance and interest. It is well known that additional stretching of the material in the unsupported wall area may prevent wrinkling, and this can be achieved in practice by increasing the blank-holder force, but the application of excessive tensile stresses leads to failure by tearing. Hence, the blank-holder force must lie within a narrow range, above that which is necessary to suppress wrinkles, on the one hand, and below that which produces fracture, on the other. This narrow range of blank-holder force is difficult to determine. For wrinkles occurring in the central area of a stamped part with a complex shape, a workable range of blank-holder force does not even exist.

The blank holder, however, does not hold the edges of the blank rigidly in place. If this were the case, tearing could occur in the cup wall. The blank holder allows the blank to slide somewhat by providing frictional force between the blank holder and the blank itself. The blank-holder force can be applied hydraulically with pressure feedback, by using an air or nitrogen cushion, or a numerically controlled hydraulic cushion. The greater the die cavity depth, the more the blank material that has to be pulled down into the die cavity and the greater the risk of wrinkling in the walls and flange of the part. The maximum die cavity depth is a balance between the onset of wrinkling and the onset of fracture, neither of which is desirable. The radii degrees of the punch and die cavity edges control the flow of blank material into the die cavity. Wrinkling in the cup wall can occur if the radii of the punch and die cavity edges are too large. If the radii are too small, the blank is prone to tearing because of the high stresses.

Effects of various factors that lead to the wrinkling apparition are the blank-holding force of the blank, the geometrical parameters of the die, the frictions that appear during deep drawing between the blank and the work elements of the die, the material characteristics and anisotropy, the contact conditions, the part geometry, the mechanical properties of the material, the imperfections in the structure, and the initial state of internal tensions of the material.

A recent trend in the car industry to use very thin high-strength steel sheets creates a situation where defects like folding and wrinkling are observed more often during stamping. The only feasible method capable of predicting such defects, before the real stamping operation takes place, is the finite element method (FEM). FEM simulation of industrial stamping processes requires in-depth knowledge of the stamping technology and, what is rarely addressed by companies offering commercial software, a good understanding of the FEM background. Frequently, the lack of experience in one of these two fields leads to unexpected results. This is especially important in the case

of simulation of wrinkling phenomena, which are closely related to sophis-
ticated problems of instability, imperfections, boundary conditions, and so
on. Wrinkling that occurs during deep drawing of conical cups was analyzed
numerically and experimentally by M. Kawka et al. The simple component
geometry enabled the easy comparison of experimental and FEM results. A
conical cup using rigid tools was selected for this investigation. A hydraulic
press with O-frame design was used for this process. The press was equipped
with a monitoring system, which enabled the acquisition of all the process-
related information. The main advantage of the hydraulic press is that the
load and ram velocity can be adjusted to the requirements of the deformation process. A similar method of rigid stops was used to control a blank
holder, that is, to maintain a constant gap between the blank holder and the
die during the drawing process. Two types of materials that are commonly
used today for deep drawing of automobile body parts were investigated:
SPCC and SPCD. The clearance between the blank holder and the blank,
resulting from the spacer height, sheet thickness, and the blank-holder force,
was equal to 0.15 mm. The aim of subsequent geometrical measurements
was to define the contours of wrinkles developed in the conical part of the
deep-drawn cup.

To compare experimental findings with numerical predictions, a 170 mm
blank divided into 4364 elements forming a fine regular mesh was used. The
material simulated was SPCE. In terms of the number of wrinkles, ABAQUS/
Explicit performed well with its 12 wrinkles as compared to 11 wrinkles found
in a cup. The amplitude of some wrinkles calculated by ABAQUS/Explicit
is comparable to that found experimentally, except for the large wrinkles
hugging the rolling direction and the transverse direction. The number of
wrinkles calculated by ITAS3D is only 8 and the amplitude is greater than the
experimental one. Research efforts on the prediction of wrinkling have been
made in the past 50 years. The analytical solution can provide a global view
in terms of the general tendency and the effect of individual parameters on
the onset of wrinkling, and can be achieved in an almost negligible computational time. However, past analytical work has been concentrated on some
relatively simple problems such as a column under axial loading, a circular
ring under inward tension, and an annular plate under bending with a conical punch at the center. Plastic bifurcation analysis is one of the most widely
used analytical approaches to predict the onset of wrinkling. Hutchinson and
Neale and Neale and Tugcu studied the bifurcation phenomenon of a doubly
curved sheet metal by adopting Donnell–Mushtari–Vlasov shell approxima-
tions. The investigation was applicable to the regions of the sheet that are free
of any surface contact.

Tugcu and coworkers extended their approach to the wrinkling of a flat plate
with infinite curvatures. Wang et al. used a similar approach to study wall
wrinkling for an anisotropic shell and applied the criterion to axisymmetric
shrink flanging. However, all these analyses are limited to long-wavelength
shallow mode and the boundary conditions or continuity conditions along

the edge of the region being examined for wrinkling are neglected. Other than the analytical approach, experiments and numerical simulations have been conducted to determine wrinkle formation tendencies in sheet metal forming. Cup-forming tests with various geometries are common experiments to investigate the side-wall wrinkling phenomenon. Yoshida et al. developed a simple test (Yoshida buckling test) to provide a reference of the wrinkle-resistant properties for various sheet metals. It involves the stretching of a square sheet along one of its diagonals. The onset and growth of wrinkles and the effects of material properties in the Yoshida test were studied analytically and numerically by Tomita and Shindo.

Numerical simulation using FEM with either an implicit or explicit integration method has become a prime tool to predict the buckling behavior in the sheet metal operation which involves complicated geometry and boundary conditions including friction. Using an implicit method to predict wrinkling is essentially an Eigen value approach, and it is hard to initiate wrinkles without initial imperfections, for example, a specific mode shape and/or material imperfection, built into the original mesh. Unlike the implicit solver, the explicit method as a dynamic approach can automatically generate deformed shapes with wrinkles due to the accumulation of numerical error. However, the onset and growth of the buckling obtained from the explicit code is sensitive to the input parameters in the FEM model, such as element type, mesh density, and simulation speed. Generally, three types of elements are employed in the sheet metal–forming simulation: membrane element, continuum element, and shell element. Membrane elements have been widely used to model the forming processes, due to their simplicity and lower computation time, especially in the inverse and optimization analysis, where many iterations of forming are required. However, it does not include bending stiffness, and therefore may not be appropriate in modeling processes where the buckling phenomenon is important, unless some special treatment (such as postprocessing) is applied.

Two kinds of experimental dies containing class "A" surface features of auto body outer panel are designed and manufactured by Jinwu Wangl and Ping Hu et al. in order to induce the surface wrinkle of an auto body panel. By conducting a stamping experiment in the dies, the wrinkling phenomenon may be observed. At the same time, numerical simulation based on an independently developed commercial CAE software KMAS/UFT of sheet metal forming is carried out. Comparison between experimental and simulated results is done, which shows good consistency. Then a universal formability theory (UFT) is introduced to optimize the adjusted amount of metal flow and the stamping dies. Some suggestions are given by investigating the adjusted amount and modification of the stamping dies.

Fuh-Kuo Chen and Yeu-Ching Liaothe studied the wrinkling problem which occurred in a stamping part of a motorcycle oil tank using the 3D finite element analysis. The material flow due to the original die design was first examined and the possible reasons causing the wrinkling problem were then identified based on the finite element analysis. In order to eliminate the wrinkles, the effects of the process parameters such as blank-holder force, blank size, and draw-bead locations on the formation of wrinkling were studied and possible solutions were tried. However, the wrinkling problem could not be improved by changing the process parameters alone. A detailed investigation of the material flow during wrinkle formation revealed that the uneven stretch between the highest portion of the part and the draw-wall at the edge of the part caused the wrinkling problem. After releasing the corner angle and cutting off some portion of the sharp corner, a sound part without wrinkles was achieved and the optimum die shape was determined. The modified die design for eliminating wrinkles is validated by the defect-free production part. The good agreement between the simulation results and the measured data obtained from the defect-free production part confirms the accuracy of the finite element analysis.

Anupam Agrawal, N. Venkata Reddy, and P.M. Dixit attempted to predict the minimum blank-holding pressure required to avoid wrinkling in the flange region during an axisymmetric deep-drawing process. Thickness variation during drawing is estimated using an upper bound analysis. The minimum blank-holding pressure required to avoid wrinkling at each punch increment is obtained by equating the energy responsible for wrinkling to the energy required to suppress the wrinkles. The predictions of the developed model are validated with the published numerical and experimental results and are found to be in good agreement. A parametric study is then conducted to study the influence of some process variables on the blank-holding pressure to avoid wrinkling.

3.3 Springback

The research work done by H. Naceur et al. deals with the optimization of deep-drawing parameters in order to reduce the springback effects after forming. A response surface method based on diffuse approximation is used; this technique has been shown to be more efficient than classical gradient-based methods since it requires fewer iterations and convergence is guaranteed especially for nonlinear problems. A new modified version of the inverse approach used to analyze the stamping operation based on DKT12 shell element is presented. In the new version, bending and unbending strains and stresses are calculated analytically from the final work piece, especially in the region of the die entrance radii, to predict the change in curvature. The bending/unbending moments and the final shape are used to calculate springback using a second incremental approach based on the updated Lagrangian formulation. The benchmark of the 2D draw-bending problem in NUMISHEET'93 has been utilized to validate the method, and good results of springback elimination have been obtained. The final results are validated using STAMPACK and ABAQUS commercial codes.

A new method for designing general sheet-forming dies to produce a desired final part shape, taking springback into account, has been developed by Wei Gan. The method is general, in that it is not limited to operations having particular symmetry, die shapes, or magnitude of springback shape change. It is based on iteratively comparing a target part shape with a finite element–simulated part shape following forming and springback. The displacement vectors at each node are used to adjust the trial die design until the target part shape is achieved; hence the term "displacement adjustment method" (DA) has been applied. DA has been compared with the "spring forward" method of Karafillis and Boyce, which is based on computing the constraint forces to maintain equilibrium following forming.

With the drive toward implementing advanced high-strength steels (AHSS) in the automotive industry, stamping engineers need to quickly answer questions about forming these strong materials into elaborate shapes. Commercially available codes have been successfully used to accurately predict formability, thickness, and strains in complex parts. However, springback and twisting are still challenging subjects in numerical simulations of AHSS components. Design of experiments has been used in the work done by A. Asgari et al. to study the sensitivity of the implicit and explicit numerical results with respect to certain arrays of user input parameters in the forming of an AHSS component. Numerical results were compared to experimental measurements of the parts stamped in an industrial production line. The forming predictions of the implicit and explicit codes were in good agreement with the experimental measurements for the conventional steel grade, while lower accuracy was observed for the springback predictions. The forming predictions of the complex component with an AHSS material were also in good correlation with the respective experimental measurements. However, much lower accuracies were observed in the springback predictions.

Springback is a very important factor that influences the quality of sheet metal forming. Accurate prediction and controlling of springback is essential for the design of tools for sheet metal forming. Wenjuan Liu et al. proposed a technique based on artificial neural network (ANN) and genetic algorithm (GA) to solve the problem of springback. An improved genetic algorithm was used to optimize the weights of the neural network. Based on the production experiment, the prediction model of springback was developed by using the integrated neural network genetic algorithm. The results show that more accurate prediction of springback can be acquired with the GA-ANN model.

3.4 Fracture

Press shops endure many metal-forming failures. Examples are die cracking, bearings running dry, press jammed on BDC, and punches wearing out prematurely. There are many others that are not so prominent. The matter of concern is sensitive failure found in most press shops that is sheet metal fracture or tearing caused by excessive stretching. Two common forms of sheet metal fracture are brittle and ductile. However, brittle fracture in sheet metal forming is uncommon. It happens for specific metal chemistries when subjected to high-impact loading at very low ($-40°$) in-service temperatures. The usual mode of a stamping fracture is ductile fracture. The amount of deformation or strain that a material can withstand before ductile fracture is very difficult, if not impossible, to predict. Microstructure, grain size, inclusions, stress state, constraints, forming speed, and many other factors control the onset of a ductile fracture. A local neck is the failure mechanism that terminates global stamping deformation. Local necking is defined as a narrow line of highly localized thinning with deformation across the neck but no deformation along the line of thinning.

Cracks on the external surface may form due to excessive tensile loads or friction. Internal cracks may form due to the presence of voids, second-phase particles, and so on. Necking during tensile deformation may result in the formation of voids, which may grow in size during loading. Cracks result due to excessive growth of voids and their coalescence. In compressive loading, surface cracks are generally formed due to excessive tensile stresses.

Bibliography

Agrawal, A., N. Venkata Reddy, and P. M. Dixit, Determination of optimum process parameters for wrinkle free products in deep drawing process, *Journal of Materials Processing Technology* 191 (2007), 51–54.

Asgari, A., M. Pereira, B. Rolfe, M. Dingle, and P. Hodgson, Design of experiments and springback prediction for AHSS automotive components with complex geometry, *API Conference Proceedings*, 778(1) (2005), 215.

Cao, J., A. Karafilis, and M. Ostrowski, Prediction of flange wrinkles in deep drawing, in: *Advanced Methods in Material Processing Defects* (Editors M. Predeleanu and P. Gilormini), Elsevier, pp. 301–310, 1997.

Chen, F.-K. and Y.-C. Liao, An analysis of draw-wall wrinkling in a stamping die design, *International Journal of Advanced Manufacturing Technology* 19 (2002), 253–259.

Chen, F.-K. and Y.-C. Liao, Analysis of draw-wall wrinkling in the stamping of a motorcycle oil tank, *Journal of Materials Processing Technology* 192–193 (2007), 200–203.

Gan, W. and R. H. Wagoner, Die design method for sheet springback, *International Journal of Mechanical Sciences* 46 (2004), 1097–1113.

Hutchinson, J. W. and K. W. Neale, Wrinkling of curved thin sheet metal, *Plastic Instability*, Presses Ponts et ChausseHes, Paris, France, 1985, pp. 71–78.

Kawka, M., L. Olejnik, A. Rosochowski, H. Sunaga, and A. Makinouchi, Simulation of wrinkling in sheet metal forming, *Journal of Materials Processing Technology* 109 (2001), 283–289.

Liu, W., Q. Liu, F. Ruana, Z. Liang, and H. Qiu, Springback prediction for sheet metal forming based on GA-ANN technology, *Journal of Materials Processing Technology* 187–188 (2007), 227–231.

Naceur, H., S. Ben-Elechi, and J. L. Batoz, The inverse approach for the design of sheet metal forming parameters to control springback effects, *European Congress on Computational Methods in Applied Sciences and Engineering (ECCOMAS)* 2004 (July 2004), 24–28.

Neale, K. W. and P. Tugcu, A numerical analysis of wrinkle formation tendencies in sheet metals, *International Journal for Numerical Methods in Engineering* 30 (1990), 1595–1608.

Ravindra Reddy, P. V. R., B. V. S. Rao, G. Chandra Mohan Reddy, P. Radhakrishna Prasad, and G. Krishna Mohan Rao, Parametric studies on wrinkling and fracture limits in deep drawing of cylindrical cup, *International Journal of Emerging Technology and Advanced Engineering* 2(6) (June 2012), 218–222, ISSN 2250-2459.

Tomita, Y. and A. Shindo, Onset and growth of wrinkles in thin square plates subjected to diagonal tension, *International Journal of Mechanical Sciences* 30 (1988), 921–931.

Tugcu, P., On plastic buckling prediction, *International Journal of Mechanical Sciences* 33 (1991), 529–539.

Venkat Reddy, R., T. A. Janardhan Reddy, and G. C. M. Reddy, Effect of various parameters on the wrinkling in deep drawing cylindrical cups, *International Journal of Engineering Trends and Technology* 3(1) (2012), 53–58.

Wang, C. T., G. Kinzel, and T. Altan, Wrinkling criterion for an anisotropic shell with compound curvatures in sheet forming, *International Journal of Mechanical Sciences* 36 (1994), 945–960.

Wang, J., P. Hu, Z. C. Fu, and Y. Xu, Experiment and numerical simulation of auto panel surface wrinkle based on universal formability technology, *Proceedings of the 2009 IEEE International Conference on Mechatronics and Automation*, Changchun, China, August 9–12, 2009.

Wang, X. and J. Cao, On the prediction of side-wall wrinkling in sheet metal forming processes, *International Journal of Mechanical Sciences* 42 (2000), 2369–2394.

Yoshida, K., J. Hayashi, and M. Hirata, Yoshida "Buckling Test", IDDRG, Kyoto, Japan, 1981.

4

Engineering Optimization

4.1 Optimization

Optimization is the act of obtaining the best results under the given circumstances. In design, construction, and maintenance of any engineering system, engineers have to take many technological and managerial decisions at several stages. The ultimate goal of all such decisions is to either minimize the efforts required or maximize the desired benefits. Since the effort required or the benefit desired in any practical situation can be expressed as a function of certain decision variables, optimization can be defined as the process of finding the conditions that give the maximum or minimum value of a function. Since the 1960s, the design of mechanical components has been greatly enhanced with the development of numerical methods. Finite element software, for instance, is now commonly used in aeronautical, mechanical, naval, and civil engineering. At the same time, efficient and fast optimization algorithms have arisen for solving various kinds of mathematical programming problems. Both trends gave birth to structural optimization, which aims at finding the best-fitted structures by modifying geometrical, material, and/or topological parameters (the variables), the optimal solution being defined with respect to at least one criterion (the objective), and having to satisfy a set of requirements (the constraints). Structural optimization has been traditionally classified into three families following the nature of the variables involved:

- In design or sizing optimization, variables represent only cross-sectional dimensions or transverse thicknesses (the geometry and the topology remaining fixed).
- In shape optimization, the variables are parameters acting directly on the geometry of the structure (but with a fixed topology).
- Finally, topological optimization handles variables that can modify the shape and the topology of the structure.

These categories are briefly illustrated in the following sections.

4.1.1 Design Optimization

In the first approach—also known as "automatic dimensioning of structures"—the only variables are cross-sectional dimensions or transverse thicknesses (the geometry and the topology remaining fixed). In trusses, for instance, the areas of the cross sections of the rods play the role of design variables, while the objective is commonly to find the lightest structure, which still satisfies a set of constraints (e.g., stresses and displacements must not overstep maximum levels).

4.1.2 Shape Optimization

In shape optimization, the variables are geometrical parameters defining the shape of the structure (the topology remaining fixed). In most works available in the literature, the parameters are the coordinates of specific points: the poles. In 2D, these poles define the contour of the structure as a set of curves, for instance, by using Lagrangian, Bezier, or B-splines interpolations. The geometry can also be modeled directly via lengths of segments, radii, angles, and so on, hence considered the design variables.

4.1.3 Topological Optimization

In topological optimization, the aim is to determine the optimal shape of a structure by starting with a bulk of material, and progressively taking off the material, which undergoes less loading. Of course, the final structure must still satisfy the user-defined constraints (generally related to the restriction of the maximum von Mises stress).

4.2 Classification of Optimization Problems

Optimization problems are classified based on various criteria:

- The nature of the variable sets: a variable may be continuous (e.g., a geometrical dimension), discrete (e.g., cross sections of beams are often available by discrete steps in catalogues), or integer (e.g., the number of layers in a composite material). There are often mixed variables in engineering problems.
- The nature of the constraints and the objective functions: they may be linear, quadratic, nonlinear, or even nondifferentiable. For instance, gradient-based algorithms, based on the computation of the sensitivities, require the functions to be differentiable in order to compute their first-order (and sometimes also their second-order) derivatives.

- The analytical properties of the functions, for example, linearity in linear programming. Convexity or monotonicity can also be successfully exploited to converge to a global optimal solution.

- The presence (or absence) of constraints. Equality constraints are usually tackled by converting them into inequality constraints.

- The size of the problem: to remain applicable even when the number of variables is very large (more than about 10,000 for continuous problems), optimization algorithms have to be adapted, because of limited memory or computational time.

- Implicit or explicit functions: in shape optimization, for instance, when finite element models are needed to compute the stresses and displacements, the objective function (generally the mass) is almost always an implicit function of the variables. Therefore, the objective(s) and constraints are approximated to a linear, quadratic, or other (cubic, posynomial, etc.) model. Neural networks may also be used to construct an approximation of the functions.

- Local or global optimization: in single-objective optimization, local optimization is used commonly with smooth functions in order to find a local optimum.

- Single-objective or multi-objective: though the first studies in structural optimization used only one objective (most of the time minimizing the mass), an increasing number of studies deal with multiple criteria (mass, cost, specific performances, etc.). Indeed, in the industrial context, optimal solutions must be good compromises between the different (and often contradictory) criteria.

To solve optimization problems, a large number of methods have been proposed in the literature. These are briefly summarized in the following section.

4.3 Local Search Methods

Local methods are aimed at reaching a local optimum, and offer no guarantee of finding the global one. The most common local methods are based on the computation of sensitivities. Therefore, they require the functions to be differentiable, and the variables to be continuous (or discrete). Nocedal et al. divided the main gradient-based algorithms into two approaches. In the *line search* strategy, the algorithm chooses a direction, for example, the steepest descent direction, and performs a search for a better point along this direction. In trust region methods, the gradient $\nabla f(x_k)$ at iteration k (and sometimes also the Hessian matrix $B(x_k)$, i.e., the second derivatives) is used to construct a model mk whose behavior is a good approximation of the original function f, at least in a close neighborhood of xk.

Various instances of these techniques have been proposed in the literature such as the conjugated gradient method and sequential quadratic programming (SQP). Though these algorithms were initially restricted to continuous unconstrained problems, they have been successfully extended to other fields of optimization.

- *Constrained optimization*: Lagrange multipliers allow the user to take constraints (equalities and inequalities) into account.
- *Discrete optimization*: In this approach, the problem is solved in a dual space to deal with discrete variables. Some interesting applications in structural optimization (sizing of thin-walled structures, geometrical configuration of trusses, topological optimization of membranes or 3D structures, etc.) can be solved.

Unfortunately, as mentioned earlier, all algorithms based on the classical gradient techniques require the functions to be differentiable, which is often not the case in design optimization problems. Furthermore, even if the functions were differentiable, the risk of being trapped in a local minimum is high. Therefore, global methods seem more suited to solve general design optimization problems. The main global approaches are summarized in the next section.

4.4 Global Search Methods

Reaching the global optimum is an arduous task, which explains that various techniques have been proposed to handle it. The most popular ones are briefly described hereafter.

4.4.1 Random Search

In this basic (and time-consuming) technique, a large number of points $x_1, x_2, ..., x_N$ are randomly generated and their corresponding function values $f(x_1), f(x_2), ..., f(x_N)$ computed; the point x^* endowed with the best function value is selected to be the solution.

4.4.2 Approximation Methods

Instead of searching directly the optimum of the true function, an approximated function is built in order to solve the problem more easily. This approximation can be a statistical function or a response surface built upon a set of function values computed for a predefined sample of variables (by the design of experiments technique). Neural network methods may also be used to approximate the objective function(s) and constraints.

4.4.3 Clustering Methods

These can be viewed as a modified form of the standard multistart procedure, which performs a local search from several points distributed over the entire search domain. A drawback of pure multistart is that when many starting points are used, the same local minimum may be obtained several times, thereby leading to an inefficient global search. Clustering methods attempt to avoid this inefficiency by carefully selecting points at which the local search is initiated.

4.4.4 Tabu Search

In this technique, with each iteration, a feasible move is applied to the current point, accepting the neighbor with the smallest cost. Tabu search acts like a local search method, except that positions that seem not favorable may be prevented from converging to the same (may be local) optimum. Tabu search also forbids reverse moves to avoid cycling (the forbidden movements are "quarantined" and composes the so-called Tabu list.

4.5 Developments in Optimization Approaches

The most commonly encountered mathematical problem in engineering is optimization. Optimization actually means to find out the best possible and desirable solution. Optimization problems have a wide range of applications and are difficult to solve, hence methods for solving such type of problems ought to be an important research topic. Optimization algorithms are of two types: deterministic and stochastic. Earlier methods that were used to solve optimization problems required massive efforts for computational solving, but these tend to fail as there is an increase in the problem size. This motivates the employment of bioinspired optimization algorithms as computationally efficient alternatives to a deterministic approach. Metaheuristic approaches are based on the iterative enhancement of a population of solutions or a single solution.

Optimization problems mapped from the real world are really challenging to solve, and many applications have to deal with nondeterministic polynomial [NP] hard problems. Therefore, for solving such problems, some special tools have to be used, though there is no assurance that the optimal solution can be found for the specified problem. In reality, there are no efficient algorithms to solve NP problems. Many problems need to be solved by trial and error by using several optimization techniques. Hence, in addition, new algorithms are developed to see whether they can solve challenging problems. Among the various new algorithms, particle swarm optimization, cuckoo search, and firefly algorithm have gained popularity due to their high efficiency in finding the optimal solution.

It can be stated that the main source of inspiration of most algorithms is from nature or biology. Therefore, almost all new algorithms can be referred to as nature inspired or bioinspired. Until now the majority of algorithms that are nature inspired were based on some characteristics of a biological system. Therefore, it can be referred to as the largest fraction of nature-inspired algorithms tending to biology-inspired, or in short bioinspired. A special class of bioinspired algorithms have been developed from observations of swarm intelligence. Algorithms that are based on swarm intelligence are most popular. Some examples are ant colony optimization, particle swarm optimization, cuckoo search, bat algorithm, and firefly algorithm.

The most recognized nature-inspired models of calculation are commonly known as cellular automata, neural computation, and evolutionary computation. More recent computational systems opted from natural processes consist of swarm intelligence, artificial immune systems, membrane computing, and amorphous computing. All the major methods and algorithms are inspired from nature and are hence known as nature-inspired metaheuristic algorithms.

4.5.1 Evolutionary Computation

Evolutionary computation is a computational model inspired by the Darwinian theory of evolution. An artificial evolutionary system is a computational system predicated on the notion of simulated evolution. It contains a constant or variable size of a population of individuals, a fitness criterion, and genetically inspired operators that engender the next generation from the current one. The initial population is typically engendered desultorily or heuristically, and typical operators are individuals or a combination of mutation and recombination. At each and every step, the individuals are evaluated according to the given fitness function (survival of the fittest). The next generation is obtained from culled individuals (parents) by utilizing genetically inspired operators. The cull of parents can be guided by a cull operator, which reflects the biological principle of mate cull. This tends to a process of simulated evolution, which eventually converges toward a proximately optimal population of individuals, from the perspective of the fitness function.

4.5.2 Swarm Intelligence

Swarm intelligence is often referred to as collective astuteness, which may also be defined as the quandary-solving behavior that instantaneously emerges from the interaction of individual agents (e.g., bacteria, ants, termites, bees, spiders, fish, birds) that communicate with other agents by action of their local environments. Particle swarm optimization applies this conception to finding an optimal solution to a given quandary by a search through a (multidimensional) solution space. At the initial stage of the swarm intelligence, a swarm of particles is defined, where each particle represents a possible solution to the quandary. Each particle has its own velocity, which depends on its precedent velocity (the inertia component), the propensity toward the past personal best position (the nostalgia component), and its propensity toward an ecumenical neighborhood optimum or local neighborhood optimum (the gregarious component). Particles thus travel through a space that is multidimensional in nature and eventually converge toward a point that is in between the ecumenical best and their personal best. Particle swarm optimization algorithms have been applied to diverse optimization quandaries, and to unsupervised learning, game learning, and scheduling applications.

There are many such types of algorithms that derive their inspirational source from various areas. Some of these are explained here with their inspirational sources and a short introduction of their workings, which are categorized into types.

4.5.3 Differential Evolution

In technology, differential evolution (DE) may be a technique that optimizes a retardant by iteratively attempting to boost a candidate answer with relevancy to a given live quality. Such ways are unremarkably referred to as metaheuristic as they create a few or no assumptions concerning the matter being optimized and might search terribly massive areas of candidate solutions.

DE is employed for 3D real-valued functions, however, without using the gradient of the matter being optimized.

4.6 Bioinspired Algorithms

Bioinspired optimization techniques have many advantages over traditional techniques and they are gaining popularity among researchers. A few such techniques are briefly discussed here and some of them are applied for sheet metal–forming optimization in the following chapters.

4.6.1 Genetic Algorithm

The genetic algorithm is a computerized search and optimization method based on the mechanics of natural genetics and natural selection. Professor John Holland of the University of Michigan, Ann Arbor, envisaged the concept of these algorithms in the mid-1960s. The genetic algorithm combines the concept of survival of the fittest among string structures with a structured yet randomized information exchange, with some of the innovative flair of human research. In every generation, a new set of artificial creatures is created using bits and pieces of the fittest; an occasional new part is tried for good measure. The genetic algorithm employs a form of simulated evolution to solve difficult optimization problems.

4.6.2 Ant Colony Optimization

Ant colony optimization (ACO) is a paradigm for designing metaheuristic algorithms for combinatorial optimization problems. ACO is a class of algorithms in which the first member, called the ant system, was initially proposed by Colorni, Dorigo, and Maniezzo. The main underlying idea, loosely inspired by the behavior of real ants, is that of a parallel search over several constructive computational threads based on local problem data and on a dynamic memory structure containing information on the quality of previously obtained results. The collective behavior emerging from the interaction of the different search threads has proved effective in solving combinatorial optimization (CO) problems.

4.6.3 Particle Swarm Optimization

Particle swarm optimization (PSO) is a new branch of soft computing paradigms called evolutionary algorithms. PSO incorporates a stable population whose individuals cooperate to find optimal solutions to difficult problems. Particle swarms also model cultural evolution. These emerge from earlier experiments with algorithms that modeled the flocking behavior seen in many species of birds. Although there were a number of such algorithms drawing considerable attention at the time, Kennedy and Eberhart became particularly interested in the models developed by biologist Frank Heppner. Heppner's "birds" exhibited most of the same flocking behaviors that other methods were producing but he added something different. His birds were attracted to a roosting area. In simulations, they would begin by flying around with no particular destination but in spontaneously formed flocks until one of the birds flew over the roosting area. When the programmed desire to roost was set higher than the desire to stay in the flock, the bird would pull away from the flock and land.

4.6.4 Artificial Bee Colony

Artificial bee colony (ABC) is a powerful metaheuristic rule introduced by Karaboga in 2005, which simulates the foraging behavior of honey bees. The alphabet rule has three phases: utilized bee, looker bee, and scout bee. In the utilized bee and looker bee phases, bees exploit the sources by native searches within the neighborhood of the solutions chosen, which are supported by the settled choice within the utilized bee part and the probabilistic choice within the looker bee part. Within the scout bee part, which is the associated analogy of abandoning exhausted food sources within the foraging method, solutions that don't seem to be helpful any longer for the search progress are abandoned, and new solutions are inserted rather than exploring new regions within the search area. The rule contains a well-balanced exploration and exploitation ability.

4.6.5 Fish Swarm Algorithm

The fish swarm algorithm (FSA) comes from the schooling behavior of fish. Cheng et al. applied the FSA for cluster analysis. The algorithmic rule operates by mimicking three vital behaviors of natural fish: looking-out behavior (tendency of fish to appear at food), swarming behavior (fish assemble in swarms to reduce danger), and following behavior (when a fish identifies food supply, its neighbors follow this fish, that is, visual power).

4.6.6 Flower Pollination Algorithm

Flowering plants have been evolving for more than 125 million years and flowers have become so influential in evolution that we cannot imagine how the plant world would be without flowers. The main purpose of a flower is ultimately reproduction via pollination. Flower pollination is typically associated with the transfer of pollen, and such transfer is often linked with pollinators such as insects, birds, bats, and other animals. Pollination can take two major forms: abiotic and biotic. About 90% of flowering plants belong to the biotic pollination type, that is, pollen is transferred by a pollinator such as insects and animals. About 10% of pollination takes the abiotic form which does not require any pollinators. Pollination can be achieved by self-pollination or cross-pollination. Cross-pollination, or allogamy, means pollination can occur from the pollen of a flower of a different plant, while self-pollination is the fertilization of one flower, such as peach flowers, from the pollen of the same flower or different flowers of the same plant, which often occurs when there is no reliable pollinator available. The flower pollination algorithm (FPA) works on the principle of pollination.

4.6.7 Cat Swarm Optimization

Cat swarm optimization (CSO) is one of the new swarm intelligence algorithms for finding the best global solution. The CSO algorithm models the behavior of cats into two modes: "seeking mode" and "tracing mode." A swarm is made up of the initial population composed of particles to search in the solution space. Here, in CSO, we use cats as particles for solving the problems. In CSO, every cat has its own position composed of D dimensions, velocities for each dimension, a fitness value that represents the accommodation of the cat to the fitness function, and a flag to identify whether the cat is in seeking mode or tracing mode. The final solution would be the best position of one of the cats. The CSO keeps the best solution until it reaches the end of the iteration.

4.6.8 Simulated Annealing

Annealing is a process in metallurgy where metals are slowly cooled to make them reach a state of low energy where they are very strong. Simulated annealing is an analogous method for optimization. It is typically described in terms of thermodynamics. The random movement corresponds to high temperature; at low temperature, there is little randomness. Simulated annealing is a process where the temperature is reduced slowly, starting from a random search at high temperature and eventually assuming a purely greedy descent as it approaches zero temperature. The randomness should tend to jump out of local minima and find regions that have a low heuristic value; greedy descent will lead to local minima. At high temperatures, worsening steps are more likely than at lower temperatures. Simulated annealing maintains a current assignment of values to variables. At each step, it picks a variable at random, and then picks a value at random. If assigning that value to the variable is an improvement or if it does not increase the number of conflicts, the algorithm accepts the assignment and there is a new current assignment. Otherwise, it accepts the assignment with some probability, depending on the temperature and how much worse it can get than the current assignment. If the change is not accepted, the current assignment is unchanged.

4.6.9 Teaching–Learning-Based Optimization

The teaching–learning-based optimization (TLBO) algorithm is a global optimization method originally developed by Rao et al. It is a population-based iterative learning algorithm that exhibits some common characteristics with other evolutionary computation (EC) algorithms. However, TLBO searches for an optimum through each learner trying to achieve the experience of the teacher, who is treated as the most learned person in the society, thereby obtaining the optimum results, rather than through learners undergoing genetic operations like selection, crossover, and mutation. Due to its simple concept and high efficiency, TLBO has become a very attractive optimization technique and has been successfully applied in many real-world problems.

4.6.10 Bacterial Foraging Optimization Algorithm

The bacterial foraging optimization algorithm (BFOA), proposed by Kevin Passino (2002), is a newcomer to the family of nature-inspired optimization algorithms. Application of the group foraging tactic of a swarm of *Escherichia coli* bacteria in multi-optimal function optimization is the main idea of this new algorithm. Bacteria search for nutrients as a way to maximize energy acquired per unit time. An individual bacterium also communicates with others by sending signals. A bacterium makes foraging choices after considering two previous factors. The process in which a bacterium searches for nutrients by moving with small steps is called chemotaxis. The main idea of BFOA is imitating the chemotactic movement of virtual bacteria in the problem search space.

4.6.11 Intelligent Water Drop Algorithm

In nature, water drops are observed in a flowing river, which form huge moving swarms. In other words, the path followed by a natural river is created by a swarm of water drops. For a swarm of water drops, the river in which they flow is the part of the environment that has been intensely changed by the swarm and will also be changed in the future. Moreover, the paths that the water drops follow are substantially affected by the environment itself. Mostly, naturally flowing rivers have paths full of twists and turns. It is believed that water drops in a river have no eyes, so without using any eyes, they can find their destination, which is obviously a lake or a sea. If we put ourselves in the place of a water drop in a flowing river, we will be able to feel that some force pulls us toward itself, which is the earth's gravity. This gravitational force has a tendency to pull everything toward the center of the earth in a straight line. Therefore, with no obstacles and barriers, the water drops should follow a straight path toward the destination, which is the shortest path from the source of the water drops to their destination (lake or ocean), which could probably be defined ideally as the earth's center. But, in reality, there are different kinds of obstacles and constraints in this path, such that the real path is very different from the ideal path. It is presumed that each water drop in a river can also carry some amount of soil. Therefore, the water drop has the ability to transfer an amount of soil from one place to another. This soil is generally transferred from fast parts to the slow parts of the path. Intelligent water drop algorithm works on this principle.

4.6.12 Bat Algorithm

Bats are very interesting animals. They are the only mammals with wings and they also possess an advanced ability of echolocation. It is estimated that there are up to 996 different species, which account for up to 20% of all mammal species. Micro bats are typically 2.2–11 cm long. Most micro bats eat insects. Micro bats use a type of sonar called echolocation for various activities such as to detect prey, to avoid obstacles, and to locate their roosting crevices in the dark. These bats emit a very loud sound pulse and listen for the echo that bounces back from the surrounding objects. Depending on the species their pulses vary in properties and can be correlated with their hunting strategies. These bats often use constant-frequency signals for echolocation but most other bats use short-frequency, modulated signals that sweep across an octave. Their signal bandwidth differs depending on the species, and often increases by the use of more harmonics. The bat algorithm mimics these principles for optimization.

4.6.13 Bee Swarm Optimization

The bee swarm optimization algorithm is inspired from the foraging behaviors of honey bee swarms. This approach uses various types of bees to find the optimal numerical functions. Each type of bee employs a different moving pattern. The scout bees fly randomly over the nearby regions. The onlooker bee selects an experienced forager bee as its motivating elite and moves toward it. An experienced forager bee memorizes the data about the best food source it has found so far, selects the best experienced forager bee as the elite bee, and adjusts its location based on this information. A colony of honey bees is capable of performing difficult tasks using relatively simple rules of a single bee's behaviors. Collecting, processing, and advertising of nectars are examples of intellectual behaviors of honey bees. A bee may be of any one type: (employed/unemployed), onlooker, forager, scout, recruit, or experienced forager. The type of a bee depends on the action which it executes and the level of information which it may use. A potential forager will reset as unemployed forager. This type of bee has no information about the environment or the position of food sources in the environment. There are two types of unemployed foragers with different flying patterns called scout bees and onlooker bees. A scout bee flies instinctively around the hive in search for new food sources without any knowledge of the environment. While waiting in the nest, the onlooker bees process the information shared by employed foragers, and select the interesting dancers. After that, the search is started, and by using the knowledge of the selected dancer bee an unemployed forager can be recruited.

4.6.14 Water Wave Optimization

The water wave theory, which was initiated with weak, nonlinear interactions among gravity waves on the surfaces of deep water, was extended by Hasselmann, and subsequently culminated in the wave turbulence theory by Zakharov et al. In shallow coastal water, the nonlinear wave field is dominated by near-resonant quadratic interactions involving triplet soft waves. It is the important interactions of wave currents at the bottom that have inspired rich and progressive coastal wave modeling since the late 1960s. Although it has a less mature relation than the well-established deep water wave models, this optimization method is motivated by shallow water wave models. The metaheuristic named water wave optimization (WWO) borrows its idea from wave motions controlled by the interactions of wave currents at the bottom to formulate the design of search mechanisms for high-dimensional global optimization problems. Experiments on a diverse set of function optimization problems show that WWO is a competitive method along with other popular metaheuristic algorithms proposed in recent years.

4.6.15 Harmony Search Algorithm

Harmony search algorithm is a new algorithm that is conceptualized from a music performance process (say, a jazz improvisation) that involves the searching for better harmony. Just as music improvisation seeks a best state (fantastic harmony) determined by aesthetic estimation, the optimization process seeks a best state (global optimum) determined by objective function evaluation. Just as aesthetic estimation is determined by the set of pitches played by an ensemble of instruments, the function evaluation is determined by the set of values assigned for decision variables. Just as aesthetic sound quality can be improved with practice, objective function value can be improved with iteration. In the analogy between improvisation and optimization, each music player (saxophonist, double bassist, or guitarist) corresponds to each decision variable and the range of each musical instrument corresponds to the range of each variable. Just as the harmony quality is enhanced by rehearsing, the solution quality is enhanced by repetition.

4.6.16 Monkey Search Algorithm

This method simulates the mountain-climbing process of monkeys. Assume that there are many mountains in a given field (i.e., in the feasible space of the optimization problem); in order to find the highest mountaintop (i.e., find the maximal value of the objective function), monkeys will climb up from their respective positions (this action is called climb process). When a monkey reaches the top of the mountain, it is natural that it will have a look to find out whether there are other mountains around it that are higher than its present whereabouts. If yes, it will jump off the mountain from its current position (this action is called watch-jump process) and then repeat the climb process until it reaches the top of the other mountain. After several repetitions of the climb process and the watch-jump process, each monkey will find a locally maximal mountaintop around its initial point. In order to find a much higher mountain top, it is natural for each monkey to somersault to a new search domain (this action is called somersault process). After many iterations of the climb process, the watch-jump process, and the somersault process, the highest mountaintop found by the monkeys will be reported as an optimal value.

4.6.17 Wolf Search Algorithm

A new bioinspired heuristic optimization algorithm called the wolf search algorithm (WSA) is based on wolf preying behavior. WSA is different from the aforementioned bioinspired metaheuristic because it simultaneously possesses both individual local searching ability and autonomous flocking movement. In other words, each wolf in the WSA hunts independently by remembering its own trait and only merges with its peer when the peer is in a better position. In this way, long-range intercommunication among the wolves that represent the searching points for candidate solutions is eliminated because wolves are known to stalk their prey in silence. Assembly depends on visual range.

Therefore, the swarming behavior of WSA, unlike most bioinspired algorithms, is delegated to each individual wolf rather than to a single leader, as in PSO, FSA, and the firefly algorithm. Effectively, WSA functions as if there are multiple leaders swarming from multiple directions to the best solution, rather than a single flock that searches for an optimum in one direction at a time. The appearance of a hunter that corresponds to each wolf is added at random and, on meeting its hunter, each wolf jumps far out of its hunter's visual range to avoid being trapped in local optima by the algorithm's design.

4.6.18 Krill Herd Algorithm

The krill herd (KH) algorithm is based on the simulation of the herding behavior of krill individuals. The minimum distances of each individual krill from food and from highest density of the herd are considered as the objective function for the krill movement. The time-dependent position of the krill individuals is formulated by three main factors: (1) movement induced by the presence of other individuals, (2) foraging activity, and (3) random diffusion. For more precise modeling of the krill behavior, two adaptive genetic operators are added to the algorithm. The proposed method is verified using several benchmark problems commonly used in the area of optimization. The KH algorithm is capable of efficiently solving a wide range of benchmark optimization problems and outperforms other exciting algorithms.

Bibliography

Akbari, R., A. Mohammadi, and K. Ziarati, A novel bee swarm optimization algorithm for numerical function optimization, *Communications in Nonlinear Science and Numerical Simulation* 15(10) (October 2010), 3142–3155.

Antonio, M. and S. Onur, Monkey search: A novel metaheuristic search for global optimization, *AIP Conference Proceedings* 953 (March 2007), 162–173.

Cheng, Y., M. Jiang, and D. Yuan, Novel clustering algorithms based on improved artificial fish swarm algorithm, *IEEE Sixth International Conference on Fuzzy Systems and Knowledge Discovery*, Tianjin, China, August 14–16, 2009, pp. 141–145.

Dorigo, M., Optimization, learning and natural algorithms, PhD thesis, Politecnico di Milano, Milan, Italy, 1992.

Fister, I., I. Fister Jr., X.-S. Yang, and J. Brest, A comprehensive review of firefly algorithms, *Swarm and Evolutionary Computation* 13 (2013), 34–46.

Fister Jr., I., X.-S. Yang, I. Fister, J. Brest, and D. Fister, A brief review of nature-inspired algorithms for optimization, *Neural and Evolutionary Computing* 80(3) (2013), 1–7.

Gandomi, A. H. and A. H. Alavi, Krill herd: A new bio-inspired optimization algorithm, *Communications in Nonlinear Science and Numerical Simulation* 17(12) (2012), 4831–4845.

Karaboga, D. and B. Basturk, On the performance of artificial bee colony (ABC) algorithm, *Applied Soft Computing* 8(1) (2008), 687–697.

Kennedy, J. and R. Eberhart, Particle swarm optimization, *Proceedings of IEEE International Conference on Neural Networks in USA*, Washington, DC, November 27–December 1, 1995, pp. 1942–1948.

Passino, K., Bio mimicry of bacterial foraging for distributed optimization and control, *IEEE Control System Management* 22(3) (2002), 52–67.

Rao, R. V., V. J. Savsani, and D. P. Vakharia, Teaching–learning-based optimization: A novel method for constrained mechanical design optimization problems, *Computer-Aided Design* 43(3) (March 2011), 303–315.

Shah-Hosseini, H., The intelligent water drops algorithm: A nature-inspired swarm-based optimization algorithm, *International Journal of Bio-Inspired Computation* 1(1/2) (2009), 71–79.

Tang, R., S. Fong, X.-S. Yang, and S. Deb, Wolf search algorithm with ephemeral memory, *Seventh International Conference on Digital Information Management*, Macau, China, August 22–24, 2012, pp. 165–172.

Tsai, H. C. and Y. H. Lin, Modification of the fish swarm algorithm with particle swarm optimization formulation and communication behaviour, *Applied Soft Computing* 11 (2011), 5367–5374.

Wang, X. et al., *An Introduction to Harmony Search Optimization Method*, Springer Briefs in Computational Intelligence, Springer, Berlin, 2015.

Yang, X.-S., *Nature-Inspired Metaheuristic Algorithms*, Luniver Press, Frome, U.K., 2008.

Yang, X.-S., Harmony search as a metaheuristic algorithm, in: *Music-Inspired Harmony Search Algorithm: Theory and Applications* (Editor Z. W. Geem), Studies in Computational Intelligence, Springer, Berlin, Germany, Vol. 191, pp. 1–14, 2009.

Yang, X.-S., A new metaheuristic bat-inspired algorithm, in: *Nature Inspired Cooperative Strategies for Optimization (NICSO 2010)* (Editors J. R. Gonzalez, D. A. Pelta, C. Cruz, G. Terrazas, and N. Krasnogor), Vol. 284, Studies in Computational Intelligence, Springer, pp. 65–74, 2010.

Yang, X.-S., Bat algorithm: Literature review and applications, *International Journal of Bio-Inspired Computation* 5(3) (2013), 141–149.

Yang, X.-S. and S. Deb, Cuckoo search via levy flights, *IEEE World Congress on Nature and Biologically Inspired Computing (NaBIC)*, Coimbatore, India, December 9–11, 2009, pp. 210–214.

Yang, X.-S. and A. H. Gandomi, Bat algorithm: A novel approach for global engineering optimization, *Engineering Computations* 29(5) (2012), 464–483.

Zheng, Y.-J., Water wave optimization: A new nature-inspired metaheuristic, *Computers & Operations Research* 55 (March 2015), 1–11.

5

Thinning Optimization: Cohort Intelligence

5.1 Cohort Intelligence Algorithm

The cohort intelligence (CI) algorithm was proposed by Dr. A.J. Kulkarni in 2013 with motivation from self-supervised learning behavior of a candidate in a cohort. This ideology is used to solve constrained optimization problems. The cohort here designates a group of candidates that are communicating and competing with each other to achieve their own individual goal, which is integrally common to all the candidates. While working in a cohort or a group, each and every candidate attempts to improve its own behavior by perceiving the behavior of every other candidate in the same cohort. Every candidate will follow a definite behavior in the cohort, which, according to itself, may lead to improvement of its own behavior. As certain qualities comply with a unique behavior, which when a candidate follows it actually tries to adjust to the associated qualities. This will make every candidate learn from one another and help the overall cohort behavior to evolve. The cohort behavior could be considered saturated if for significant number of learning attempts the individual behavior of all the candidates does not improve considerably and candidates' behaviors become hard to differentiate.

5.2 General Constrained Optimization Problem

Consider a general constrained problem (in the minimization sense) as follows:

```
Minimize f(x) = f(x₁,…xᵢ,…x_N)
Subject to
gᵢ(x) ≤ 0, i = 1…n
hᵢ(x) = 0, i = 1…n
Ψᵢ^lower ≤ xᵢ ≤ Ψᵢ^upper, i = 1…n
```

Consider the objective function f(x) as the behavior of an individual candidate in the cohort. Every candidate consistently tries to enhance its behavior by modifying the associated sets of characteristics/attributes/qualities. $x = (x_1, ..., x_i, ..., x_N)$.

The basic steps of CI can be summarized as the pseudocode as follows:

```
Begin
Objective function f(x), x = (x1, ..., xd) T
Select number of candidates in cohort (C).
Set sampling interval reduction factor (r).
Set convergence factor (ε).
    While (t < max learning attempts) or convergence criterion,
    Initialize qualities/attributes of each candidate by
        random numbers,
    Evaluate fitness/behavior of every candidate.
    Evaluate probability associated with the behavior being
        followed by every candidate in the cohort. Use roulette
        wheel approach to select behavior to follow by each
        candidate within the cohort (C). Every candidate will
        shrink/expand the sampling interval of every quality
        with respect to the followed behavior.
If (Convergence criterion met)
    Accept the current best candidate and its behavior.
Break
End
End
```

The optimization procedure begins with the initialization of the number of candidates C, sampling interval Ψ_i for each quality x_i, $i = 1, 2, ..., N$, learning attempt counter n = 1, and setting up of static sampling interval reduction factor r $\in[0, 1]$, convergence parameter $\in$, and number of variations t.

Here we use two types of penalty function approaches:

1. Static penalty approach

   If static penalty is used, the problem formulation will become

$$f_P(X) = f(X) + \sum_{i=1}^{m} P_i \times (g_i(x))^2 + \sum_{i=1}^{m} P_i \times (h_i(x))^2$$

2. Dynamic penalty approach

   If dynamic penalty is used, the problem formulation will become

$$eval(\bar{x}) = f(\bar{x}) + (C \times t)^a \sum_{j=1}^{m} f_j^\beta(\bar{x})$$

Hence, $f^*(x^c) = f^*(x^c)$ + penalty value (Figure 5.1).

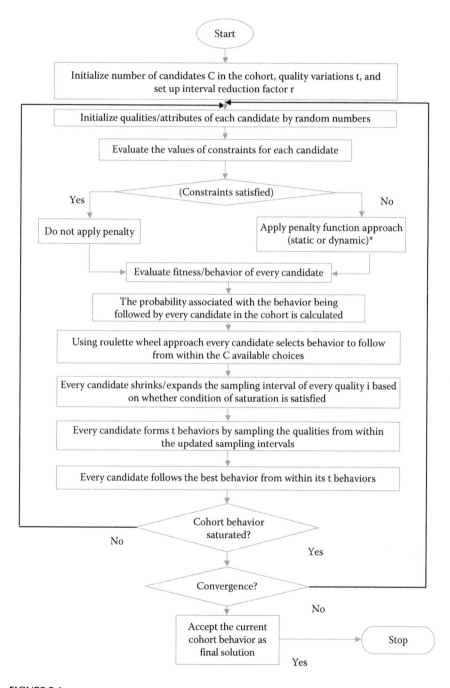

FIGURE 5.1
Cohort intelligence algorithm.

The step-by-step procedure of how the algorithm works is explained as follows:

Step 1: The probability of selecting the behavior $f^*(x^c)$ of every associated candidate c is calculated as follows:

$$p^c = \frac{1/f^*\left(x^c\right)}{\sum_{c=1}^{C} 1/f^*\left(x^c\right)}$$

Step 2: Every candidate c generates a random number $\in[0, 1]$, and using a roulette wheel approach decides to follow the corresponding behavior $f^*(x^{c[?]})$ and associated qualities $x^{c[?]} = x_1^{c[?]},\ldots,x_i^{c[?]},\ldots,x_N^{c[?]}$. The superscript indicates that the behavior is selected by candidate c and not known in advance. The roulette wheel approach could be most appropriate as it provides a chance for every behavior in the cohort to get selected purely based on its quality.

Step 3: Every candidate c (c = 1, ..., C) shrinks the sampling interval $\Psi_i^{c[?]}$, i = 1, 2, ..., N associated with every variable $x_i^{c[?]}$, i = 1, 2, ..., N to its local neighborhood. This is done as follows:

$$\Psi_i^{c[?]} = \left[x_i^{c[?]} - \frac{\|\Psi_i\|}{2}, x_i^{c[?]} + \frac{\|\Psi_i\|}{2} \right]$$

where $\Psi_i = \|\Psi_i\| \times r$.

Step 4: Every candidate c (c = 1, ..., C) samples t qualities from within the updated sampling interval $\Psi_i^{c[?]}$, i = 1, 2, ..., N associated with every quality $x_i^{c[?]}$, i = 1, 2, ..., N and computes a set of associated t behaviors, that is, $F^{c,t} = \{f(x^c)^1, \ldots, f(x^c)^j, \ldots, f(x^c)^t\}$ and selects the best behavior $f^*(x^c)$ from within. This makes the cohort available with C updated behaviors, represented as

$$F^C = \left\{ f^*\left(x^1\right),\ldots,f^*\left(x^c\right),\ldots,f^*\left(x^C\right) \right\}$$

Step 5: The cohort behavior could be considered saturated if there is no significant improvement in the behavior $f^*(x^c)$ of every candidate c (c = 1, ..., C) in the cohort, and the difference between the individual behaviors is not very significant for a considerable number of successive learning attempts, that is, if

1. $\|\max(F^C)^n - \max(F^C)^{n-1}\| \le \epsilon$
2. $\|\min(F^C)^n - \min(F^C)^{n-1}\| \le \epsilon$
3. $\|\max(F^C)^n - \min(F^C)^n\| \le \epsilon$

4. Every candidate c (c =1, ..., C) expands the sampling interval $\Psi_i^{c[?]}$, i = 1, ..., N associated with every quality $x_i^{c[?]}$, i = 1, ..., N to its original one $\Psi_i^{lower} \leq x_i \leq \Psi_i^{upper}$, i = 1, ..., n.

Step 6: If either of the two following criteria is valid, accept any of the C behaviors from the current set of behaviors in the cohort as the final objective function value f*(x) as the final solution and stop, else continue to Step 1.

1. If maximum number of learning attempts is exceeded.
2. If cohort saturates to the same behavior (satisfying the conditions in Step 5) for τ_{max} times.

5.3 Penalty Function

Based on mathematical programming approaches, a constraint numerical optimization problem is transformed into an unconstrained numerical optimization problem (pseudo-objective function), and optimization algorithms have adopted penalty functions, whose general formula is the following:

$$\varnothing(\vec{x}) = f(\vec{x}) + p(\vec{x})$$

where $\varnothing(\vec{x})$ is the expanded objective function to be optimized, and $p(\vec{x})$ is the penalty value that can be calculated as follows:

$$p(\vec{x}) = \sum_{i=1}^{m} r_i \cdot \max\left(0, g_i(\vec{x})^2\right) + \sum_{j=1}^{p} C_j \cdot |h_j(\vec{x})|$$

where r_i and c_j are positive constants called penalty factors.

The main motive of using penalty factors is to favor the selection of feasible solutions over infeasible solutions. There are two types of penalty function methods:

1. Interior penalty function or barrier method
2. Exterior penalty function method

5.3.1 Interior Penalty Function or Barrier Method

The interior penalty function method transforms any constrained optimization problem into an unconstrained one. However, the barrier functions prevent the current solution from ever leaving the feasible region. They require the interior of the feasible sets to be nonempty, which is impossible if equality constraints are present. Therefore, they are used with problems having only inequality constraints.

Commonly used penalty functions are

$$B(x) = -\sum_{j=1}^{m} \frac{1}{g_j(x)}$$

or

$$B(x) = -\sum_{j=1}^{m} \log\left(-g_j(x)\right)$$

The auxiliary function now becomes

$$f(x) + \mu B(x)$$

where μ is a small positive number.

5.3.2 Exterior Penalty Function Method

Like the interior penalty function method, this method also transforms any constrained problem into an unconstrained one. The constraints are incorporated into the objective by means of a "penalty parameter," which penalizes any constraint violation. The larger the constraint violation, the larger is the objective function penalized.

The penalty function p(x) is defined as follows:

$$p(x) = \sum_{j=1}^{m} \left[\max\{0, g_j(x)\} \right]^{\alpha} + \sum_{j=1}^{m} \left| h_j(x) \right|^{\alpha}$$

5.4 Static Penalty Function Method

A simple method to penalize infeasible solutions is to apply a constant penalty to those solutions that violate feasibility in any way. The penalized objective function would then be the unpenalized objective function plus a penalty (for a minimization problem). A variation is to construct this simple penalty function as a function of the number of constraints violated where there are multiple constraints. The penalty function for a problem with m constraints would then be

$$f_P(X) = f(X) + \sum_{i=1}^{m} C_i \times \delta_i$$

where
$\delta_i = 1$, if constraint i is violated
$\delta_i = 0$, if constraint i is satisfied
$f_p(X)$ is the penalized objective function
$f(X)$ is the unpenalized objective function
C_i is a constant imposed for violation of constraint i

The following is a general formulation for a minimization problem:

$$f_P(X) = f(X) + \sum_{i=1}^{m} P_i \times (g_i(x))^2 + \sum_{i=1}^{m} P_i \times (h_i(x))^2$$

where
$g(X)$ is the inequality constraints
$h(X)$ is the equality constraints

5.5 Dynamic Penalty Function

In this constraint-handling technique, the individuals are evaluated based on the following formula:

$$eval(\overline{x}) = f(\overline{x}) + (C \times t)^a \sum_{j=1}^{m} f_j^\beta(\overline{x})$$

where C, α, and β are constants.

This method initially penalizes infeasible solutions and, as the iteration number increases, it heavily penalizes infeasible solutions. The penalty component in this method changes with each generation number. However, this method is very sensitive to the parameters C, α, and β, and so these parameters need to be properly tuned before this technique is applied. In this technique, the penalty components have a significant effect on the objective function as they constantly keep increasing with the generation number.

5.6 Proposed Innovative Methodology

An innovative methodology has been proposed for optimization of sheet metal–forming cylindrical geometrical components:

1. Industrial circular geometrical components encompassing almost all features/configurations have been selected for studying the drawing process.
2. Four major process parameters—blank-holder force, friction condition, punch nose radius, and die profile radius—have been selected for Taguchi design of experiments.
3. Experiments have been conducted with simulations using finite element simulation software.
4. Results of simulations have been examined in the light of performance characteristics/measures as thinning, thickening, wrinkling, fracture, springback, and thickness gradient. Thickness gradient is a novel criterion that has been applied for the first time.
5. Linear relationships have been established between process parameters and various performance measures.
6. Optimization problem has been formulated.
7. Different bioinspired algorithms have been applied for optimization using MATLAB®.
8. The results of optimization have been validated with experimental results.

The same methodology has been applied in all coming chapters.

5.7 Configurations of Components

The ultimate aim of the study was to optimize the metal-forming process for cylindrical geometrical automotive components. It was necessary to include components with almost all configurations used in the automotive sector. Five different components were selected for the study. All cylindrical configurations used in automotive components can be observed in the selected five components. These components were again classified according to their configurations. The components under study contain the following features:

- Circular top and circular bottom cylindrical components
- Circular top and elliptical bottom components

- Components with and without flange
- Components with straight walls and inclined walls
- Single step and multistep components
- Components with plain circular bottom
- Components with holes in circular bottom
- Components with holes and features in circular/elliptical bottom

The configurations and dimensions of the components under study are described in the following chapters.

5.8 Sealing Cover

The sealing cover is manufactured by Vishwadeep Enterprises, Pune. It is fitted to two-wheeler petrol tanks. The configuration of the sealing cover is very simple, but quite different from that of a typical cup. It has a large diameter-to-height ratio and there is no flange. The base has a curvature with 2 mm radius. The cup has a diameter of 58.6 mm and the total height is 9 mm (Figure 5.2).

The other details are provided in Table 5.1.

FIGURE 5.2
Sealing cover.

TABLE 5.1

Sealing Cover Details

Manufactured by	Vishwadeep Enterprises, Pune
Part No.	J-019
Weight	28 g
Material	D-513, SS 4010, UST 1203
Thickness	0.8 mm
Yield strength	192 MPa
Ultimate tensile strength	315 MPa
R	1.7 min
N	0.22
Material model	Anisotropic

5.9 Selection of Process Parameters and Performance Measure: Thinning

Based on a literature survey and industrial insights, four major process parameters have been selected for investigation:

1. Blank-holder force
2. Coefficient of friction
3. Punch nose radius
4. Die profile radius

For the component under study, thinning is selected as a performance measure. Experiments are designed to study and correlate the effect of process parameters on thinning.

5.10 Numerical Investigations: Taguchi Design of Experiments

Nine experiments designed as per L9 orthogonal array have been conducted using finite element simulation software, and the results of experimentation have been used for analyzing the performance parameters. The following are the various types of results achieved for thickness:

- Thickness
- Thickness strain (engineering %)

TABLE 5.2

Sealing Cover—Three Levels of Process Parameters

	Lower	Middle	Higher
BHF [kN]	02	03	04
μ	0.05	0.10	0.15
R_D [mm]	1.5	2.0	2.5
R_P [mm]	6.0	7.0	8.0

TABLE 5.3

Sealing Cover—L9 Orthogonal Array

Experiment No.	Blank-Holder Force [kN]	Coefficient of Friction	Die Profile Radius [mm]	Punch Nose Radius [mm]
1	02	0.05	1.5	6.0
2	02	0.10	2.0	7.0
3	02	0.15	2.5	8.0
4	03	0.05	2.0	8.0
5	03	0.10	2.5	6.0
6	03	0.15	1.5	7.0
7	04	0.05	2.5	7.0
8	04	0.10	1.5	8.0
9	04	0.15	2.0	6.0

Every process variable has three levels of operation: low, medium, and high. The orthogonal array selected for this combination of four parameters and three levels is L9. Table 5.2 provides the parameters and their three levels. Table 5.3 presents the detailed L9 orthogonal array.

The original thickness of the sealing cover is 0.8 mm. It is observed during experimentation that there is thinning as well as thickening behavior. The thickness ranges for nine experiments are presented in Table 5.4. Figures 5.3 through 5.11 show the thickness distribution in all nine experiments.

TABLE 5.4

Sealing Cover—Thickness Range

Experiment 1	0.78–0.84 mm	Experiment 2	0.77–0.84 mm	Experiment 3	0.77–0.84 mm
Experiment 4	0.78–0.84 mm	Experiment 5	0.77–0.84 mm	Experiment 6	0.76–0.84 mm
Experiment 7	0.78–0.84 mm	Experiment 8	0.77–0.84 mm	Experiment 9	0.76–0.84 mm

Thickness (mm)

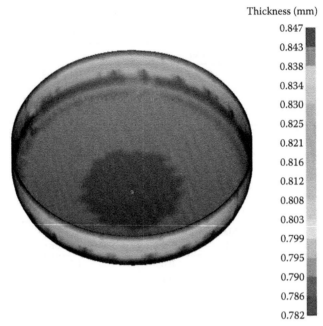

FIGURE 5.3
Sealing cover—thickness distribution in experiment 1.

Thickness (mm)

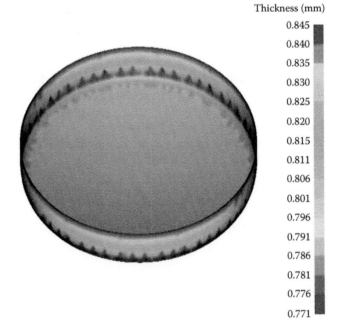

FIGURE 5.4
Sealing cover—thickness distribution in experiment 2.

Thickness (mm)

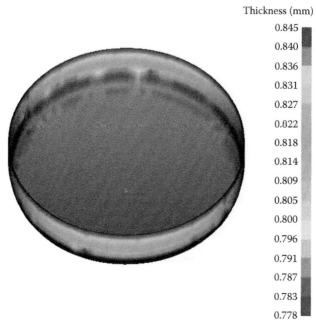

FIGURE 5.5
Sealing cover—thickness distribution in experiment 3.

Thickness (mm)

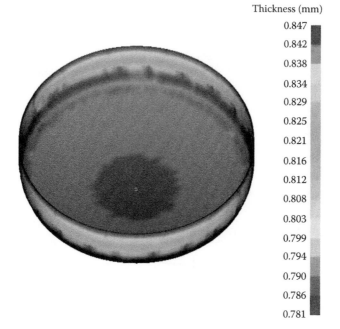

FIGURE 5.6
Sealing cover—thickness distribution in experiment 4.

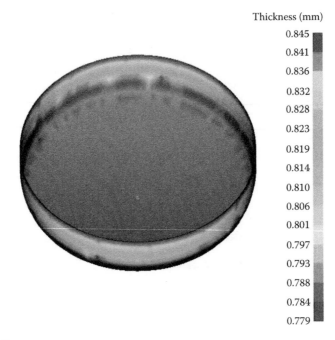

FIGURE 5.7
Sealing cover—thickness distribution in experiment 5.

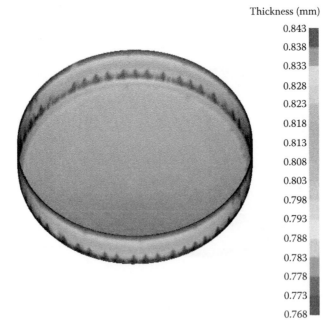

FIGURE 5.8
Sealing cover—thickness distribution in experiment 6.

Thickness (mm)

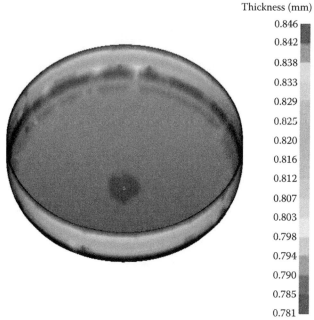

FIGURE 5.9
Sealing cover—thickness distribution in experiment 7.

Thickness (mm)

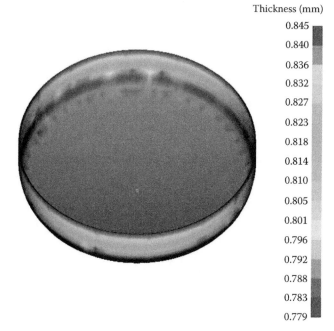

FIGURE 5.10
Sealing cover—thickness distribution in experiment 8.

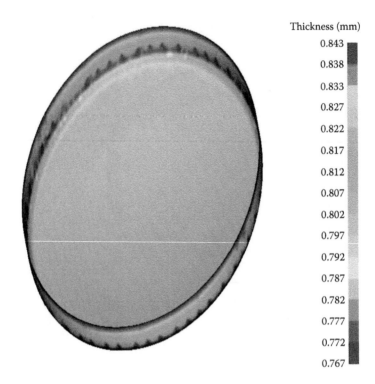

Thickness (mm)

0.843	
0.838	
0.833	
0.827	
0.822	
0.817	
0.812	
0.807	
0.802	
0.797	
0.792	
0.787	
0.782	
0.777	
0.772	
0.767	

FIGURE 5.11
Sealing cover—thickness distribution in experiment 9.

5.11 Analysis of Variance

The decrease in thickness at various cross sections during all nine experiments is measured and presented in Table 5.5. The thickness is measured at various cross sections and the average thickness is calculated. For analysis of variance, the quality characteristic selected is the difference between the original thickness and the decreased thickness. The S/N ratios are calculated for the quality characteristic, where the smaller the ratio, the better is the quality.

The minimum thickness observed is 0.77 mm. The average decrease in thickness for experiment 4 is the minimum observed, which is 0.005 mm. The results of the analysis of variance are presented in Tables 5.6 and 5.7. The mean S/N ratios are calculated for all parameters—blank-holder force, coefficient of friction, die profile radius, and punch nose radius—at all three levels, that is, low, medium, and high. The range is defined as the difference between the maximum and minimum values of the S/N ratio for a particular parameter.

Table 5.7 shows the rearrangement of S/N ratios for all variables at all levels. The rank indicates the influence of the input parameter on the quality

TABLE 5.5

Sealing Cover—Thickness Distribution

Experiment No.	Decreased Thickness [mm]			Average Thickness [mm]	Thickness Difference [mm]	S/N Ratio
1	0.784	0.793	0.797	0.791	0.008	41.24
2	0.793	0.789	0.773	0.785	0.015	36.47
3	0.780	0.790	0.794	0.788	0.012	38.41
4	0.792	0.797	—	0.794	0.005	45.19
5	0.782	0.790	0.795	0.789	0.011	39.17
6	0.790	0.780	0.786	0.786	0.013	37.39
7	0.792	0.796	0.787	0.791	0.008	41.58
8	0.794	0.790	—	0.792	0.008	41.93
9	0.795	0.790	0.785	0.790	0.010	40.00

TABLE 5.6

Sealing Cover—S/N Ratios at Three Levels for Thinning

Parameter	Level	Experiments	Mean S/N Ratio
Blank-holder force (BHF)	1	1, 2, 3	38.71
	2	4, 5, 6	40.58
	3	7, 8, 9	41.17
Coefficient of friction (μ)	1	1, 4, 7	42.67
	2	2, 5, 8	39.19
	3	3, 6, 9	38.60
Die profile radius (R_D)	1	1, 6, 8	40.19
	2	2, 4, 9	40.55
	3	3, 5, 7	39.72
Punch nose radius (R_P)	1	1, 5, 9	40.13
	2	2, 6, 7	38.48
	3	3, 4, 8	41.73

TABLE 5.7

Sealing Cover—Analysis of Variance (ANOVA) Results for Thinning

	BHF	Friction	R_D	R_P
1	38.71	42.67	40.19	40.13
2	40.58	39.19	40.55	38.48
3	41.17	38.60	39.72	41.73
Range	2.46	4.06	0.832	3.24
Rank	3	1	4	2

characteristic. The result for this orthogonal array indicates that friction has a major influence on the decrease in thickness. The punch nose radius is second in rank, the blank-holder force is the third, and the die profile radius has the least influence on thickness reduction.

5.12 Objective Function Formulation

Linear mathematical relations have been developed from the results of Taguchi design of experiments and analysis of variance between input parameters like blank-holder force, friction coefficient, die profile radius, and punch nose radius. The performance characteristic applied for sealing cover is thinning. The relationships are presented as follows. Minitab has been used for regression analysis. No fracture has been observed during experimentation.

The objective function is

Minimize thinning

$$\text{Thinning} = 0.782 + (0.000778 * \text{BHF}) - (0.0433 * \mu)$$
$$- (0.00039 * R_D) + (0.00069 * R_P)$$

Subject to

$$1.2 \leq \beta \leq 2.2$$
$$3R_D \leq R_P \leq 6R_D$$
$$F_{d\ max} \leq \pi d_m S_0 S_u$$
$$R_D \geq 0.035 \left[50 + (d_0 - d_1) \right] \sqrt{S_0}$$

5.13 Results

The CI algorithm was applied for optimization of thinning. The parameters selected for the algorithm are provided in Table 5.8.

During the minimization process, the diameter of the sealing cover and process variables such as die profile radius and coefficient of friction were selected as variables. Table 5.9 presents the results of optimization and optimized value of variables with the lower and upper limits.

The optimum value of thickness achieved is 0.778 mm. So thinning happens up to the extent of 0.022 mm. The optimum diameter obtained is 60.96 mm. The coefficient friction needs to be maintained at an optimum with a value of 0.14. The die profile radius obtained is 2.43 mm. Figure 5.12 indicates the graph of optimum values obtained with every iteration of the CI algorithm.

TABLE 5.8

Cohort Parameters

Cohort Parameters	Set Value
Cohort population	5
Interval reduction factor	0.96
Number of iterations	2000

TABLE 5.9

Optimization Results—Cohort Intelligence

Parameter	Lower Bound	Upper Bound	Optimum
Sealing cover diameter [mm]	57	61	60.96
Die profile radius [mm]	1.5	2.5	2.43
Coefficient of friction	0.005	0.15	0.14
Optimized thickness			0.778 mm

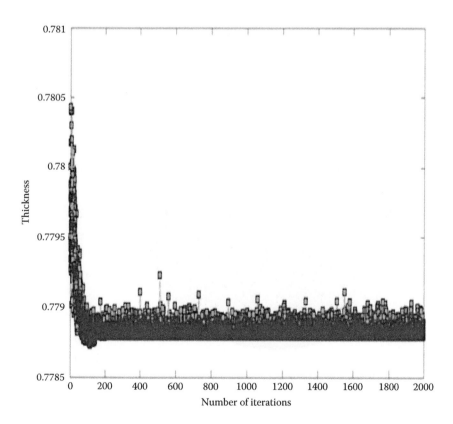

FIGURE 5.12

Optimum thickness iteration wise.

5.14 Validation: Numerical Simulation

The numerical simulation was carried out with the optimum parameters achieved after optimization applying CI. The thickness observed at different sections in the sealing cover is presented in Figures 5.13 through 5.16. The average thickness in all sections is measured and observed to be 0.78 mm. This indicates that when optimum parameters are selected, a smaller thinning value is obtained.

5.15 Experimental Validation

Despite the fact that sheet metal–forming technologies are extensively used in modern industry, the tooling and production process is still largely based on empirical results. The development of numerical simulation determines the capability of an objective assessment of formability, the strain distribution at

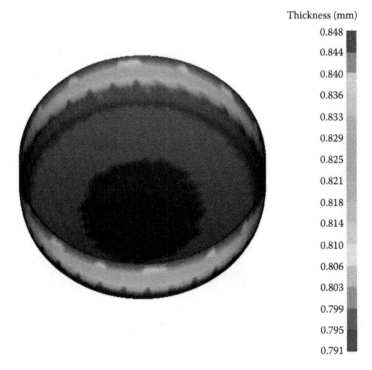

Thickness (mm)

| 0.848 |
| 0.844 |
| 0.840 |
| 0.836 |
| 0.833 |
| 0.829 |
| 0.825 |
| 0.821 |
| 0.818 |
| 0.814 |
| 0.810 |
| 0.806 |
| 0.803 |
| 0.799 |
| 0.795 |
| 0.791 |

FIGURE 5.13
Sealing cover—thickness distribution with optimized parameters.

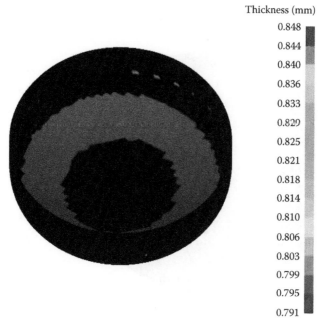

FIGURE 5.14
Sealing cover—ranges of thickness distribution with optimized parameters.

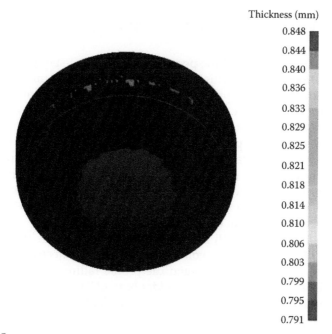

FIGURE 5.15
Sealing cover—ranges of thickness distribution with optimized parameters.

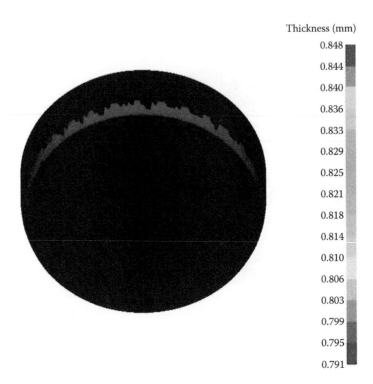

Thickness (mm)

0.848
0.844
0.840
0.836
0.833
0.829
0.825
0.821
0.818
0.814
0.810
0.806
0.803
0.799
0.795
0.791

FIGURE 5.16
Sealing cover—ranges of thickness distribution with optimized parameters.

different stages of draw, and the possibility of reducing stamping trials during the assessment process. This saves resources in terms of time, energy, and money. Whenever a new approach is developed using numerical simulations, it is necessary to validate it experimentally to certify the numerical results.

To validate the results of the new proposed methodology, the sealing cover has been selected for experimental validation. The press used for experimentation is of 100 ton capacity, is clutch-operated, and has a single acting mechanical power press of H-type. It has a steel body with a bed size of 680 × 680 mm. The bed to ram distance is 585 mm and the stoke length is 125 mm. The prime mover for the press is an electric motor of 10 hp capacity. The shaft speed is 3680 rpm (Figures 5.17 and 5.18).

For components with high diameter to depth ratio, it is observed that blank-holder force and friction condition have a more effective influence on forming than punch nose radius and die profile radius. Experiments were conducted using an optimum blank-holder force of 2.04 kN and coefficient of friction of 0.14. As it is a single acting press, the blank-holder force is applied with bottom spring adjustments. To maintain the film lubrication and coefficient of friction at 0.14, grease was used as a lubricant and the component was formed. Experimental formability analysis was carried out for manufactured

FIGURE 5.17
Mechanical press.

FIGURE 5.18
Tooling design.

components and a forming-limit diagram was plotted by observing and measuring strain distribution at various grid points. The manufactured components with optimum parameters with different configurations of a circle grid are shown in Figures 5.19 and 5.20. The forming-limit diagram doesn't show any failure point, hence concluding that forming is successful and the component is safe.

FIGURE 5.19
Different configurations of circle grids for experimentation.

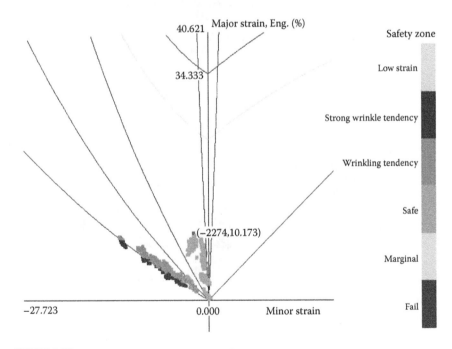

FIGURE 5.20
Forming-limit diagram—experimental results.

Bibliography

Kulkarni, A. J., I. P. Durugkar, and M. Kumar, Cohort intelligence: A self supervised learning behavior, *IEEE International Conference on Systems, Man, and Cybernetics*, Manchester, UK, 2013, pp. 1393–1400.

Kulkarni, A. J. and H. Shabir, Solving 0–1 knapsack problem using cohort intelligence algorithm, *International Journal of Machine Learning and Cybernetics* 7(3) (2014), 1–15.

Mezura-Montesa, E. and C. A. Coello, Constraint-handling in nature-inspired numerical optimization: Past, present and future, *Swarm and Evolutionary Computation* 1 (2011), 173–194.

Periyanan, P. R., U. Natarajan, and S. H. Yang, A study on the machining parameters optimization of micro-end milling process, *International Journal of Engineering, Science and Technology* 3(6) (2011), 237–246.

6

Springback Optimization: Flower Pollination

6.1 Introduction

Nature has been solving challenging problems over millions and billions of years, and many biological systems have evolved with intriguing and surprising efficiency in maximizing their evolutionary objectives such as reproduction. Based on the successful characteristics of biological systems, many nature-inspired algorithms have been developed over the last few decades. For example, the genetic algorithm (GA) is an optimization method that mimics the principle of natural genetics and natural selection to constitute search and optimization procedures. The population is evolved using the operators such as selection, crossover, and mutation. GA often gives optimized results and performs well under various circumstances. Similar to GA, the approach of differential evolution (DE) is mutation-driven, which helps explore and further locally exploit the solution space to reach the global optimum. Although easy to implement, there are several problem-dependent parameters that need to be fine-tuned and the procedure may also require several associated trials to be performed. Particle swarm optimization (PSO) is a population-based stochastic optimization technique developed by Dr. Eberhart and Dr. Kennedy in 1995, inspired by the social behavior of bird flocking or fish schooling. It simulates the movement of animals as in a flock of birds or a school of fish. The PSO concept consists of changing the velocity of (accelerating) each particle, at each time step, toward its p_{best} and l_{best} locations (local version of PSO). Cuckoo search (CS) is an optimization algorithm developed by Xin-She Yang and Suash Deb in 2009. It was inspired by the obligate brood parasitism of some cuckoo species that lay their eggs in the nests of other host birds (of other species). The gray wolf optimizer (GWO) algorithm mimics the leadership hierarchy and hunting mechanism of gray wolves in nature.

Flower pollen localization algorithm is a nature-inspired algorithm which simulates the pollination of flowering plants. Pollination means transfer of pollen from one flower to another flower of the same or another plant. This transfer of pollen can happen through pollinators such as birds, insects, bats, and other animals, which is termed biotic pollination. Pollination by wind and diffusion is called abiotic pollination. Further, pollination can be classified into cross-pollination and self-pollination. If pollination or fertilization happens between flowers of different flower plants, it is called cross-pollination. If pollination happens between flowers of the same flowering plant, it is called self-pollination. Pollination of flowers is a process of reproduction and survival of the fittest of a particular plant species. This fitness characteristic is used to define the optimization problem.

6.2 Flower Pollination

Flower pollen localization algorithm is a nature-inspired algorithm which derives its idea from the characteristics of flowering plants. This localization algorithm is referred from the flower algorithm developed by Xin-She Yang in 2012. On the earth, 80% of plants are flowering plants. Flowers are used for reproduction of their own species through pollination. Pollination means the transfer of pollen from one flower to another flower in the same plant or another plant. This transfer of pollen can happen through pollinators such as birds, insects, bats, and other animals.

Pollination by birds, animals, bats, and insects is termed biotic pollination. Pollination by wind and diffusion is called abiotic pollination. Further, pollination can be classified into cross-pollination and self-pollination. If pollination or fertilization happens between flowers of different flower plants, it is called cross-pollination. If pollination happens between flowers of the same flowering plant, it is called self-pollination. Pollination of flowers is a process of reproduction and survival of the fittest of a particular plant species. This fitness characteristic is used to define the optimization of the localization problem in wireless sensor networks. Some insects and bees have Lévy weight behavior, meaning their jumps or flying steps obey a Lévy distribution. Some pollinators have developed flower constancy, which means that some flowers are related to some birds and insects. They are interdependent. Therefore, particular birds and insects jump or fly only to certain species of flower plants. And those flowers also provide the food required by those particular birds and insects. This flower constancy increases the pollination process in specific flower species, thus maximizing reproduction.

The flower pollen localization algorithm uses the following rules:

Rule 1: Biotic and cross-pollination can be considered to be a global pollination process, and pollen-carrying pollinators move in a way that obeys Lévy flights.

Rule 2: For local pollination, biotic and self-pollination are used.

Rule 3: Pollinators such as insects can develop flower constancy, which is equivalent to a reproduction probability that is proportional to the similarity of two flowers involved.

Rule 4: The interaction or switching of local pollination and global pollination can be controlled by a switch probability $p \in (1, 0)$ with a slight bias toward local pollination.

The rules can be formulated into updating equations as

$$x_i^{t+1} = x_i^t + \gamma L(\lambda)\left(x_i^t - B\right)$$

where
x_i^t is the pollen i or solution vector x_i at iteration t
B is the current best solution found among all solutions at the current generation/iteration
γ is the scaling factor to control the step size
$L(\lambda)$ is the parameter that corresponds to the strength of the pollination

Since insects may move over a long distance with steps of various lengths, we can use a Lévy flight to imitate this characteristic efficiently. To model local pollination, both rule 2 and rule 3 can be represented as

$$x_i^{t+1} = x_i^t + U\left(x_j^t - x_k^t\right)$$

where x_j^t and x_k^t are pollen from different flowers of the same plant species. This essentially imitates the flower constancy in a limited neighborhood. For rule 4, a switch probability value can be set (usually 0.8) to switch between common global pollination and intensive local pollination.

The selection of the flower pollination algorithm (FPA) was based on its capability to converge fast compared to other nature-inspired algorithms, as well as its compatibility with other algorithms such as PSO, CS algorithm, firefly algorithm, and GWO, which allows it to be hybridized. This capability of FPA is important as the optimization problem is rather complex and selection of an improper algorithm may lead to faulty design.

6.3 Flower Pollination—Flowchart

Flower pollination algorithm is explained in the flowchart shown in Figure 6.1.

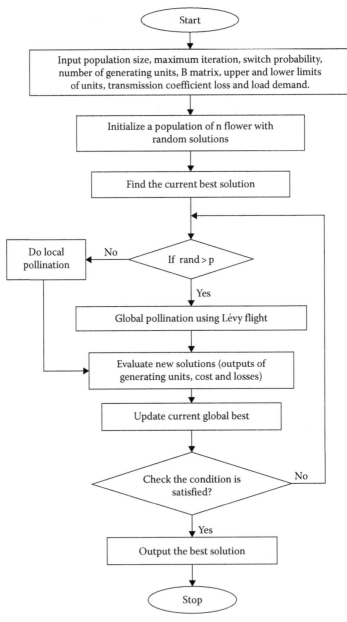

FIGURE 6.1
Flower pollination algorithm.

6.4 Pseudocode

```
Define Objective function f (x), x = (x1, x2, ..., xd)
Initialize a population of n flowers/pollen gametes with
  random solutions
Find the best solution B in the initial population
Define a switch probability p ∈ [0, 1]
Define a stopping criterion (either a fixed number of
  generations/iterations or accuracy)
While (t <Max Generation)
for i = 1 : n (all n flowers in the population)
if rand <p,
Draw a (d-dimensional) step vector L which obeys a Lévy
  distribution
Global pollination via xᵢᵗ⁺¹ = xᵢᵗ+L(B-xᵢᵗ)
else
Draw U from a uniform distribution in [0, 1]
Do local pollination via xᵢᵗ⁺¹ = xᵢᵗ+U(xⱼᵗ-xₖᵗ)
end if
Evaluate new solutions
If new solutions are better, update them in the population
end for
Find the current best solution B
end while
```

Output the best solution found.
The FPA has the following steps:

Step 1: Initialize N flowers from solution space.

Step 2: Find the objective function of each flower and select global flower from the population.

Step 3: For each flower generate a random number.

Step 4: If the generated random number is less than the switching probability, go to step 6.

Step 5: Select a flower randomly from the population and use it for local pollination, then go to step 7.

Local pollination is represented by the equation

$$x_i^{t+1} = x_i^t + U\left(x_j^t - x_k^t\right)$$

Step 6: Pollinate a flower with global flower.

Global pollination is represented by the equation

$$x_i^{t+1} = x_i^t + \gamma L(\lambda)(x_i^t - B)$$

L is the Lévy weight.

Step 7: Till N number of flower, repeat step 3 to step 6.

Step 8: Till maximum number of generation, repeat step 2 to step 7.

Step 9: Print best value of the flower among the generations.

6.5 Sealing Cover

The sealing cover is manufactured by Vishwadeep Enterprises, Pune. It is fitted to two-wheeler petrol tanks. The configuration of a sealing cover is very simple, but quite different from that of a typical cup. It has a large diameter to height ratio and there is no flange. The base has a curvature of 2 mm radius. The cup has a diameter of 58.6 mm and its total height is 9 mm (Figure 6.2).

The other details are provided in Table 6.1.

FIGURE 6.2
Sealing cover.

TABLE 6.1

Sealing Cover Details

Manufactured by	Vishwadeep Enterprises, Pune
Part No.	J-019
Weight	28 g
Material	D-513, SS 4010, UST 1203
Thickness	0.8 mm
Yield strength	192 MPa
Ultimate tensile strength	315 MPa
r	1.7 min
n	0.22
Material model	Anisotropic

6.6 Selection of Process Parameters and Performance Measure: Springback

Based on a literature survey and industrial insights, four major process parameters have been selected for investigation:

1. Blank-holder force
2. Coefficient of friction
3. Punch nose radius
4. Die profile radius

Springback is selected as a performance measure. Experiments are designed to study and correlate the effect of the process parameters on springback.

6.7 Numerical Investigations: Taguchi Design of Experiments

Nine experiments designed as per L9 orthogonal array have been conducted using finite element simulation software, and the results achieved after experimentation have been used for analyzing the performance parameters. The following are the various types of results achieved for springback:

- Springback displacement magnitude
- Springback displacement deflection

Every process variable has three levels of operation: low, medium, and high. The orthogonal array selected for this combination of four parameters and

TABLE 6.2

Sealing Cover—Three Levels of Process Parameters

	Lower	Middle	Higher
BHF [kN]	02	03	04
μ	0.05	0.10	0.15
R_D [mm]	1.5	2.0	2.5
R_P [mm]	6.0	7.0	8.0

three levels is L9. Table 6.2 provides the parameters and their three levels. Table 6.3 presents the detailed L9 orthogonal array.

During numerical experimentation springback is observed. Springback during all designed experiments is presented Table 6.4. and the state of springback in all nine experiments is shown in Figures 6.3 through 6.11.

In experiment 3, the springback displacement magnitude at the bottom central region and in some portion of the bottom wall is in the range of 0.033–0.040 mm. There are regions at the bottom where the magnitude is

TABLE 6.3

Sealing Cover—L9 Orthogonal Array

Experiment No.	Blank-Holder Force [kN]	Coefficient of Friction	Die Profile Radius [mm]	Punch Nose Radius [mm]
1	02	0.05	1.5	6.0
2	02	0.10	2.0	7.0
3	02	0.15	2.5	8.0
4	03	0.05	2.0	8.0
5	03	0.10	2.5	6.0
6	03	0.15	1.5	7.0
7	04	0.05	2.5	7.0
8	04	0.10	1.5	8.0
9	04	0.15	2.0	6.0

TABLE 6.4

Sealing Cover—Springback Displacement Magnitude

Experiment 1	0.002–0.108 mm	Experiment 2	0.002–0.107 mm	Experiment 3	0.005–0.109 mm
Experiment 4	0.002–0.108 mm	Experiment 5	0.004–0.108 mm	Experiment 6	0.006–0.105 mm
Experiment 7	0.003–0.108 mm	Experiment 8	0.004–0.108 mm	Experiment 9	0.007–0.105 mm

Springback displacements: magnitude (mm)

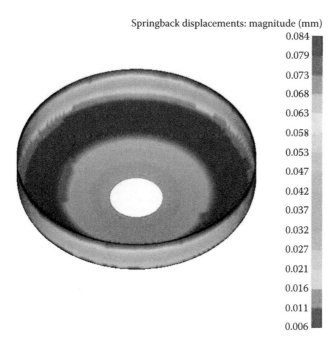

FIGURE 6.3
Sealing cover—springback displacement magnitude in experiment 1.

Springback displacements: magnitude (mm)

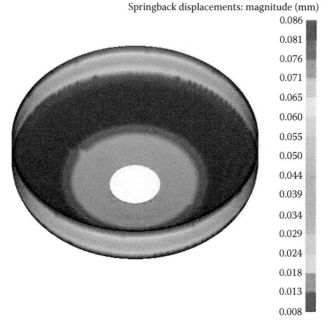

FIGURE 6.4
Sealing cover—springback displacement magnitude in experiment 2.

Springback displacements: magnitude (mm)

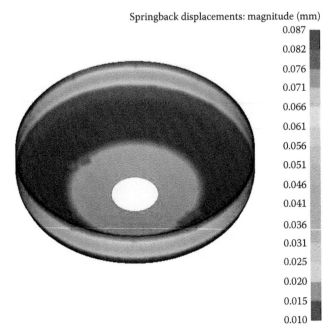

FIGURE 6.5
Sealing cover—springback displacement magnitude in experiment 3.

Springback displacements: magnitude (mm)

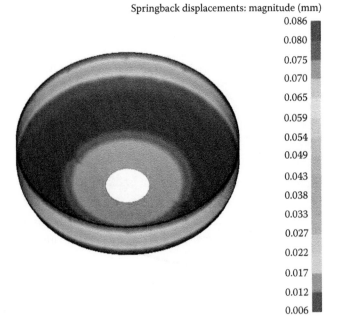

FIGURE 6.6
Sealing cover—springback displacement magnitude in experiment 4.

Springback displacements: magnitude (mm)

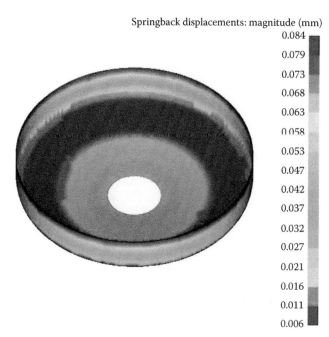

FIGURE 6.7
Sealing cover—springback displacement magnitude in experiment 5.

Springback displacements: magnitude (mm)

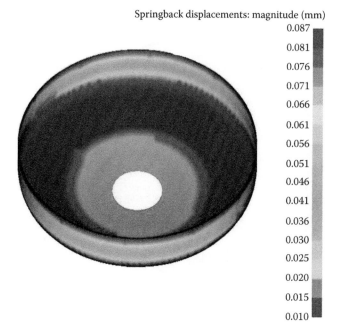

FIGURE 6.8
Sealing cover—springback displacement magnitude in experiment 6.

Springback displacements: magnitude (mm)

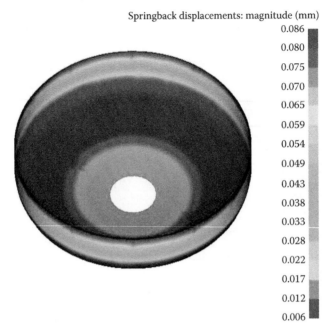

FIGURE 6.9
Sealing cover—springback displacement magnitude in experiment 7.

Springback displacements: magnitude (mm)

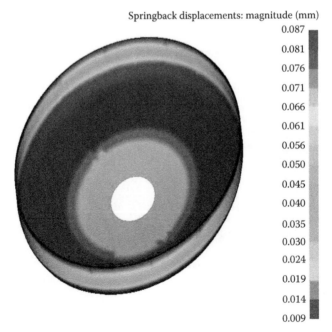

FIGURE 6.10
Sealing cover—springback displacement magnitude in experiment 8.

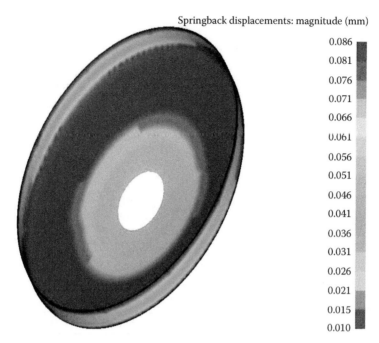

Springback displacements: magnitude (mm)

0.086	
0.081	
0.076	
0.071	
0.066	
0.061	
0.056	
0.051	
0.046	
0.041	
0.036	
0.031	
0.026	
0.021	
0.015	
0.010	

FIGURE 6.11
Sealing cover—springback displacement magnitude in experiment 9.

0.019–0.026 and 0.005–0.012 mm. The wall has a displacement magnitude ranging between 0.054 and 0.088 mm. At certain locations in the upper end of the wall, the magnitude is between 0.102 and 0.109 mm.

6.8 Analysis of Variance

The springback at various cross sections during all nine experiments is measured and presented in Table 6.5. The average springback is calculated from these values. For analysis of variance, the quality characteristic selected was springback. The S/N ratios are calculated for the quality characteristic, where the smaller the ratio, the better the quality.

The maximum springback displacement magnitude is 0.040 mm and the minimum is 0.013 mm. The maximum springback displacement, observed in experiment 8, is 0.036 mm. The results of the analysis of variance are presented in Tables 6.6 and 6.7. The mean S/N ratios are calculated for all parameters—blank-holder force, coefficient of friction, die profile radius,

TABLE 6.5

Sealing Cover—Springback

Experiment No.	Springback Displacement [mm]				Average Displacement [mm]	S/N Ratio
1	0.014	0.020	0.028	0.035	0.028	30.80
2	0.013	0.020	0.034	0.040	0.026	31.53
3	0.015	0.023	0.030	0.037	0.022	32.84
4	0.020	0.027	0.043	0.033	0.026	31.70
5	0.015	0.022	0.028	0.035	0.032	29.89
6	0.016	0.023	0.030	0.036	0.027	31.21
7	0.014	0.020	0.027	0.035	0.033	29.57
8	0.015	0.022	0.028	0.035	0.036	28.87
9	0.017	0.024	0.030	0.037	0.028	31.05

TABLE 6.6

Sealing Cover—S/N Ratios at Three Levels for Springback

Parameter	Level	Experiments	Mean S/N Ratio
Blank-holder force [BHF]	1	1, 2, 3	31.72
	2	4, 5, 6	30.93
	3	7, 8, 9	29.83
Coefficient of friction [μ]	1	1, 4, 7	30.69
	2	2, 5, 8	30.10
	3	3, 6, 9	31.70
Die profile radius [R_D]	1	1, 6, 8	30.29
	2	2, 4, 9	31.43
	3	3, 5, 7	30.77
Punch nose radius [R_P]	1	1, 5, 9	30.58
	2	2, 6, 7	30.77
	3	3, 4, 8	31.13

TABLE 6.7

Sealing Cover—ANOVA Results for Springback

	BHF	Friction	R_D	R_P
1	31.72	30.69	30.29	30.58
2	30.93	30.10	31.43	30.77
3	29.83	31.70	30.77	31.13
Range	1.89	1.60	1.13	0.55
Rank	1	2	3	4

and punch nose radius—at all three levels, that is, low, medium, and high. The range is defined as the difference between the maximum and minimum values of S/N ratio for a particular parameter.

The result for the orthogonal array indicates that the blank-holder force has a major influence on springback displacement magnitude. Friction is second in rank, the die profile radius is third, and the punch nose radius has the least influence on springback displacement.

6.9 Objective Function Formulation

Linear mathematical relations have been developed from the results of Taguchi design of experiments and analysis of variance between input parameters like blank-holder force, friction coefficient, die profile radius, and punch nose radius. The performance characteristics applied for sealing cover is springback. The relationships are presented below. Minitab has been used for regression analysis.

The objective function is

Minimize springback

$$\text{Spring back displacement magnitude} = 0.0239 + (0.00159 * \text{BHF})$$
$$- (0.0324 * \mu) - (0.00144 * R_D)$$
$$+ (0.00067 * R_P)$$

Subject to

$$1.2 \leq \beta \leq 2.2$$
$$3R_D \leq R_P \leq 6R_D$$
$$F_{d\ max} \leq \pi d_m S_0 S_u$$
$$R_D \geq 0.035 \left[50 + (d_0 - d_1) \right] \sqrt{S_0}$$

6.10 Results

Flower pollination has been applied for optimization. The parameters selected for the algorithm are presented in Table 6.8.

TABLE 6.8

Flower Pollination Parameters

Flower Pollination Parameters	Set Value
Number of iterations	2000
Population size	20
Probability switch	0.8

TABLE 6.9

Optimization Results—Flower Pollination

Parameter	Lower Bound	Upper Bound	Optimum
Sealing cover diameter [mm]	57	61	61
Die profile radius [mm]	1.5	2.5	2.49
Coefficient of friction	0.005	0.15	0.14
Optimized springback displacement magnitude			0.02 mm

During the minimization process, the diameter of the sealing cover and process variables such as die profile radius and coefficient of friction were selected as variables. Table 6.9 presents results of optimization and optimized value of variables with the lower and upper limits.

The optimum value of springback displacement magnitude achieved is 0.02 mm. The optimum diameter obtained is 61 mm. The coefficient friction needs to be maintained at an optimum value of 0.14. The die profile radius obtained is 2.49 mm.

6.11 Validation: Numerical Simulation

The numerical simulation was conducted with the optimum parameters achieved after optimization applying FPA. The thickness observed at different sections in the sealing cover is presented in Figures 6.12 through 6.15. The average springback in all sections is measured and observed to be 0.02 mm. This indicates that if optimum parameters are selected, a reduced springback is obtained.

Springback displacements: magnitude (mm)

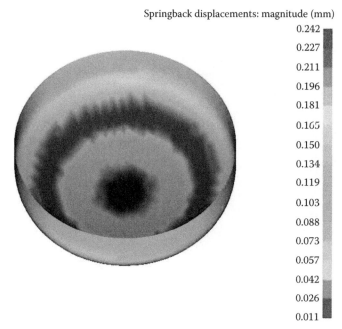

FIGURE 6.12
Sealing cover—ranges of springback displacement magnitude with optimized parameters.

Springback displacements: magnitude (mm)

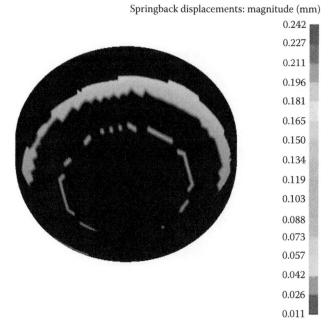

FIGURE 6.13
Sealing cover—ranges of springback displacement magnitude with optimized parameters.

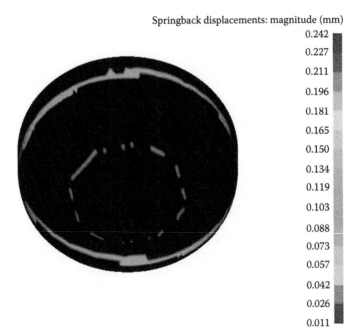

FIGURE 6.14
Sealing cover—ranges of springback displacement magnitude with optimized parameters.

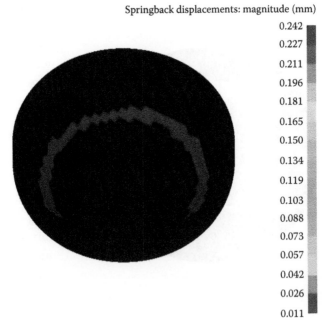

FIGURE 6.15
Sealing cover—ranges of springback displacement magnitude with optimized parameters.

Bibliography

Abdel-Raouf, O., M. Abdel-Baset, and I. El-henawy, A new hybrid flower pollination algorithm for solving constrained global optimization problems, *International Journal of Applied Operational Research* 4(2) (2014), 1–13.

Hegazy, O., O. S. Soliman, and M. A. Salam, Comparative study between FPA, BA, MCS, ABC, and PSO algorithms in training and optimizing of LS-SVM for stock market prediction, *International Journal of Advanced Computer Research* 5(18) (March 2015), 35–45.

Łukasik, S. and P. A. Kowalski, Study of flower pollination algorithm for continuous optimization, *Advances in Intelligent Systems and Computing* 322 (2015), 451–459.

Sakib, N., M. W. U. Kabir, M. S. Rahman, and M. S. Alam, A comparative study of flower pollination algorithm and bat algorithm on continuous optimization problems, *International Journal of Applied Information Systems (IJAIS)* 7(9) (2014), 13–19.

Wang, R. and Y. Zhou, Flower pollination algorithm with dimension by dimension improvement, *Mathematical Problems in Engineering* 2014 (2014), 1–9 Article ID 481791.

Yang, X.-S., Flower pollination algorithm for global optimization, *Unconventional Computation and Natural Computation 2012*, Lecture Notes in Computer Science 7445 (2012), 240–249.

7

Fracture Optimization: Genetic Algorithm Approach

7.1 Introduction

The genetic algorithm (GA) is a computerized search and optimization method based on the mechanics of natural genetics and natural selection. Professor John Holland of the University of Michigan, Ann Arbor, envisaged the concept of these algorithms in the mid-1960s. The GA combines the concept of survival of the fittest among string structures with a structured yet randomized information exchange, with some of the innovative flair of human research. In every generation, a new set of artificial creatures is created using bits and pieces of the fittest; an occasional new part is tried for good measure. The GA employs a form of simulated evolution to solve difficult optimization problems. A GA is a blind search technique, where the answer is not known and the algorithm explodes along until an answer is found in the search space, which is good enough to be used as an answer. A GA simulates the Darwinian theory of evolution using highly parallel, mathematical algorithms that transform a set (population) of mathematical objects (typically strings of 1s and 0s) into a new population, using operators such as reproduction, mutation, and crossover. The initial population is selected at random, either by the toss of a coin, generated by a computer, or by some other means, and the algorithm will continue until a certain time has elapsed or a certain condition is met. These algorithms are best suited to solve optimization problems that cannot be solved using conventional methods. Thus, they are often applied to problems that are nonlinear and with multiple local optima. Implementation of the GA begins with a population of (typically random) artificial chromosomes. These structures are then evaluated and reproductive opportunities are allocated in such a way that those chromosomes that represent a better solution to the target problem are given more chances to "reproduce" than those chromosomes representing poorer solutions. The goodness of a solution may be computed by comparing its evaluation against the population average, or it may be a function of the rank of that individual in the population relative to other solutions. The term

GA can have two meanings. In a strict interpretation, it refers to the model introduced and investigated by John Holland and his PhD students. Most of the existing theories of GA apply either solely or primarily to this model. In a broader sense, a GA is any population-based model that uses selection and recombination operators and mutation operators to generate new sample points in a search space. Evolutionary algorithms do not use gradient information. This makes it possible to use genetic and evolutionary algorithms for applications where other mathematical optimization techniques are not appropriate. These algorithms are particularly useful for ill-structured search problems that are characterized by having a large number of local optima.

7.1.1 Coding

In order to use GAs to solve a problem, variables (x_i) are first coded in some string structures. It is important to mention here that the coding of the variables is not absolutely necessary. There exist some studies where GAs are directly used on the variables themselves, but here we shall ignore the exceptions and discuss the working principles of a simple GA. Binary-coded strings with 1s and 0s are mostly used [106]. The length of the string is usually determined according to the desired accuracy of the solution. For example, if four bits are used to code each variable in a two-variable function optimization problem, the strings (0000 0000) and (1111 1111) would represent the points

$$\left(x_1^{(L)}, x_2^{(L)}\right)^T \quad \left(x_1^{(U)}, x_2^{(U)}\right)^T$$

respectively, because the substrings (0000) and (1111) have the minimum and maximum decoded values. Any other eight-bit string can be found to represent a point in the search space according to a fixed mapping rule. Usually, the following linear mapping rule is used:

$$x_i = x_i^{(L)} + \frac{x_i^{(U)} - x_i^{(L)}}{2^{l_i} - 1} \quad \text{Decoded value}(S_i)$$

In this equation, the variable x_i is coded in a substring s_i of length l_i. The decoded value of a binary substring s_i is calculated as $\sum_{i=0}^{l-1} 2^i S_i$, where $S_i \in$ (0, 1) and the string s is represented as ($S_{l-1} S_{l-2} \ldots S_2 S_1 S_0$). It is not necessary to code all the variables in equal substring length. The length of a substring representing a variable depends on the desired accuracy in that variable.

7.1.2 Fitness Function

As pointed out earlier, GAs mimic the survival-of-the-fittest principle of nature to make a search process. Therefore, GAs are naturally suitable for solving maximization problems. Minimization problems are usually transformed into maximization problems by some suitable transformation. In general, a fitness function F(x) is first derived from the objective function and used in successive genetic operations. Certain genetic operators require that the fitness function be nonnegative. For maximization problems, the fitness function can be considered to be the same as the objective function or F(x) = f(x). For minimization problems, the fitness function is an equivalent maximization problem chosen such that the optimum point remains unchanged. A number of such transformations are possible. The following fitness function is often used:

$$F(x) = \frac{1}{\left(1 + f(x)\right)}$$

This transformation does not alter the location of the minimum, but converts a minimization problem to an equivalent maximization problem. The fitness function value of a string is known as the string's fitness.

7.1.3 Reproduction

Reproduction is usually the first operator applied on a population. Reproduction selects good strings in a population and forms a mating pool. That is why the reproduction operator is sometimes known as the selection operator. There exist a number of reproduction operators in GA literature, but the essential idea is that the above-average strings are picked from the current population and their multiple copies are inserted in the mating pool in a probabilistic manner. The commonly used reproduction operator is the proportionate reproduction operator where a string is selected for the mating pool with a probability proportional to its fitness. Thus, the ith string in the population is selected with a probability proportional to F_i. Since the population size is usually kept fixed in a simple GA, the sum of the probability of each string being selected for the mating pool must be 1. Therefore, the probability for selecting the ith string is

$$p_i = \frac{F_i}{\sum_{j=1}^{n} F_i}$$

where n is the population size. One way to implement this selection scheme is to imagine a roulette wheel with its circumference marked for each string proportionate to the string's fitness. The roulette wheel is spun n times, each time selecting an instance of the string chosen by the roulette wheel pointer. Since the circumference of the wheel is marked according to a string's fitness, this roulette wheel mechanism is expected to make

$F_i / \bar{F}$ copies of the ith string in the mating pool. The average fitness of the population is calculated as

$$\bar{F} = \sum_{i=1}^{n} \frac{F_i}{n}$$

Figure 7.1 shows a roulette wheel for five individuals having different fitness values.

Using the fitness value F_i of all strings, the probability of selecting a string pi can be calculated. Thereafter, the cumulative probability (P_i) of each string being copied can be calculated by adding the individual probabilities from the top of the list. Thus, the bottom-most string in the population should have a cumulative probability (P_n) equal to 1. The roulette wheel concept can be simulated by realizing that the ith string in the population represents the cumulative probability values from P_{i-1} to P_i. The first string represents the cumulative values from zero to P1. Thus, the cumulative probability of any string lies between 0 and 1. In order to choose n strings, n random numbers between zero and one are created. Thus, a string that represents the chosen random number in the cumulative probability range (calculated from the fitness values) for the string is copied to the mating pool. This way, the string with a higher fitness value will represent a larger range in the cumulative probability values and therefore has a higher probability of being copied into the mating pool. On the other hand, a string with a smaller fitness

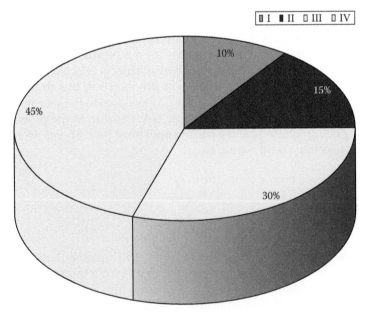

FIGURE 7.1
Example of roulette wheel used for reproduction.

value represents a smaller range in cumulative probability values and has a smaller probability of being copied into the mating pool. In reproduction, good strings in a population are probabilistically assigned a larger number of copies and a mating pool is formed. It is important to note that no new strings are formed in the reproduction phase.

7.1.4 Crossover

In the crossover phase, new strings are created by exchanging information among strings of the mating pool. Many crossover operators exist in the GA literature. In most crossover operators, two strings are picked from the mating pool at random and some portions of the strings are exchanged between the strings. A single-point crossover operator is performed by randomly choosing a crossing site along the string and by exchanging all bits on the right side of the crossing site as shown:

$$0\cdot0 | 0\cdot0\cdot0 \quad \rightarrow \quad 0\cdot0\cdot1\cdot1\cdot1$$

$$1\cdot1 | 1\cdot1\cdot1 \quad \rightarrow \quad 1\cdot1\cdot0\cdot0\cdot0$$

The two strings participating in the crossover operation are known as parent strings and the resulting strings are known as children strings. It is intuitive from this construction that good substrings from parent strings can be combined to form better children strings, if an appropriate site is chosen. Since the knowledge of an appropriate site is usually not known beforehand, a random site is often chosen. With a random site, the children strings produced may or may not have a combination of good substrings from parent strings, depending on whether or not the crossing site falls in the appropriate place. But we do not worry about this too much, because if good strings are created by crossover, there will be more copies of them in the next mating pool generated by the reproduction operator. But if good strings are not created by crossover, they will not survive too long, because reproduction will select against those strings in subsequent generations.

It is clear from this discussion that the effect of crossover may be detrimental or beneficial. Thus, in order to preserve some of the good strings that are already present in the mating pool, not all strings in the mating pool are used in crossover. When a crossover probability of p_c is used, only 100 p_c percent strings in the population are used in the crossover operation and 100 $(1 - p_c)$ percent of the population remains as they are in the current population.

7.1.5 Mutation

A crossover operator is mainly responsible for the search of new strings, even though a mutation operator is also used for this purpose sparingly. The mutation operator changes 1 to 0 and vice versa with a small mutation probability p_m.

The mutation is performed bit by bit by flipping a coin with a probability p_m. If at any bit the outcome is true, then the bit is altered; otherwise, it is kept unchanged. The need for mutation is to create a point in the neighborhood of the current point, thereby achieving a local search around the current solution. The mutation is also used to maintain diversity in the population. For example, consider the following population having four eight-bit strings:

0110	1011
0011	1101
0001	0110
0111	1100

Notice that all four strings have a 0 in the left-most bit position. If the true optimum solution requires 1 in that position, then neither reproduction nor the crossover operator described earlier will be able to create 1 in that position. The inclusion of mutation introduces some probability (N_{pm}) of turning 0 into 1.

These three operators are simple and straightforward. The reproduction operator selects good strings and the crossover operator recombines good substrings from good strings to hopefully create a better substring. Even though none of these claims is guaranteed and/or tested while creating a string, it is expected that if bad strings are created they will be eliminated by the reproduction operator in the next generation and if good strings are created, they will be increasingly emphasized.

7.2 Tail Cap

The tail cap is manufactured by Vishwadeep Enterprises, Chikhali, Pune. It is fitted at the exhaust end of a two-wheeler's silencer. The tail cap configuration is most complicated (Figure 7.2). It has a trapezoidal section with a top diameter of 110 mm and the bottom is elliptical with a major diameter of 80 mm and a minor diameter of 70 mm. The total height of the cup is 43 mm, where the upper 8 mm is the straight portion and the remaining is inclined. The corner radius between the inclined wall and the straight wall is 2.7 mm. The bottom is also inclined with a central hole of 21.5 mm diameter emerging outward. The corner radius at the bottom is 5 mm and in the radius between the inclined bottom and the hole is 2 mm. The hole has a height of 5.8 mm.

Table 7.1 presents details of a tail cap.

FIGURE 7.2
Tail cap.

TABLE 7.1

Tail Cap Details

Manufactured by	Vishwadeep Enterprises, Pune
Component of	Kinetic Motors Company Ltd.
Part No.	IKS 0024A
Weight	166 g
Material	SPCD
Thickness	1.2 mm
Yield strength	250 MPa
Ultimate tensile strength	270–370 MPa
r	1.3 min
n	0.18 min

7.3 Selection of Process Parameters and Performance Measure: Fracture

Based on a literature survey and industrial insights, four major process parameters have been selected for investigation:

1. Blank-holder force
2. Coefficient of friction
3. Punch nose radius
4. Die profile radius

Fracture has been selected as the performance measure. Experiments are designed to study and correlate the effects of process parameters on fracture.

7.4 Numerical Investigations: Taguchi Design of Experiments

Nine experiments designed as per L9 orthogonal array have been conducted using finite element simulation software, and the results achieved after experimentation have been used for analyzing the performance parameters. Fracture has been assessed using a failure limit diagram with a novel fracture factor proposed here. For analysis of variance the addition of the square of vertical distances of all points above the failure curve is considered for minimization, so as to convert all points from the failure region to a safe region to minimize fracture. The least-square method is employed for distance measurement of fracture points. The ultimate objective is to convert all failure points to a safe zone by minimizing the total distance on the failure limit diagram. The fracture factor is defined in terms of distances, which represent major strains on the failure limit diagram.

Every process variable has three levels of operation: low, medium, and high. The orthogonal array selected for this combination of four parameters and three levels is L9. Table 7.2 presents the parameters and their three levels. Table 7.3 presents the detailed L9 orthogonal array.

Forming limit diagrams have been plotted for all nine experiments (Figures 7.3 through 7.11). The failure occurs at 39.90 MPa of major strain. The maximum strain recorded on the forming limit diagram is 102.42 MPa; there are a few points at which a value above this has been recorded in all experiments. The maximum minor strain recorded is 48.91 MPa. Almost all the points are either in the safe zone or in the marginal zone. There are also points in the wrinkling zone in all experiments.

TABLE 7.2

Tail Cap—Three Levels of Process Parameters

	Lower	Middle	Higher
BHF [kN]	20	25	30
μ	0.05	0.10	0.15
R_D [mm]	1.5	2.0	2.5
R_P [mm]	6.0	7.0	8.0

TABLE 7.3

Tail Cap—L9 Orthogonal Array

Experiment No.	Blank-Holder Force [kN]	Coefficient of Friction	Die Profile Radius [mm]	Punch Nose Radius [mm]
1	20	0.05	1.5	6.0
2	20	0.10	2.0	7.0
3	20	0.15	2.5	8.0
4	25	0.05	2.0	8.0
5	25	0.10	2.5	6.0
6	25	0.15	1.5	7.0
7	30	0.05	2.5	7.0
8	30	0.10	1.5	8.0
9	36	0.15	2.0	6.0

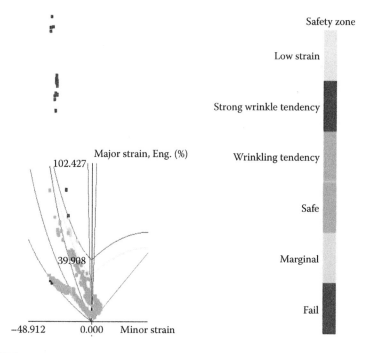

FIGURE 7.3

Tail cap—forming limit diagrams in experiment 1.

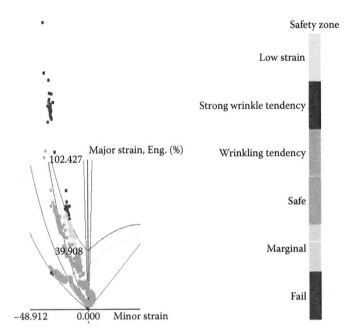

FIGURE 7.4
Tail cap—forming limit diagrams in experiment 2.

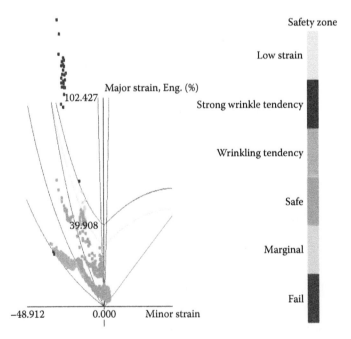

FIGURE 7.5
Tail cap—forming limit diagrams in experiment 3.

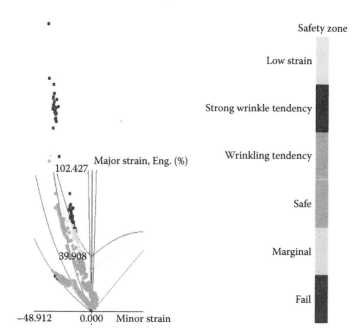

FIGURE 7.6
Tail cap—forming limit diagrams in experiment 4.

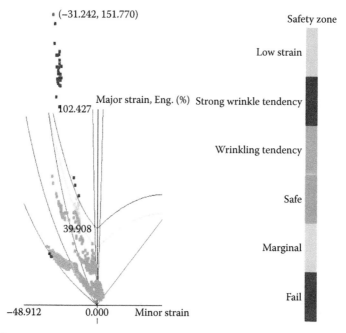

FIGURE 7.7
Tail cap—forming limit diagrams in experiment 5.

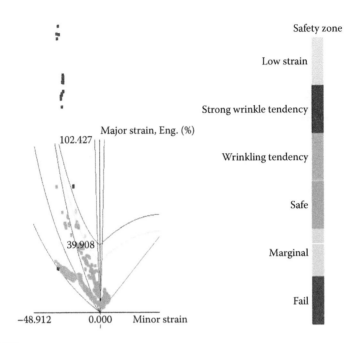

FIGURE 7.8
Tail cap—forming limit diagrams in experiment 6.

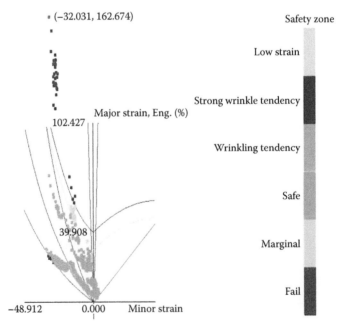

FIGURE 7.9
Tail cap—forming limit diagrams in experiment 7.

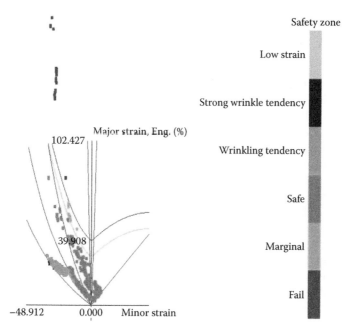

FIGURE 7.10
Tail cap—forming limit diagrams in experiment 8.

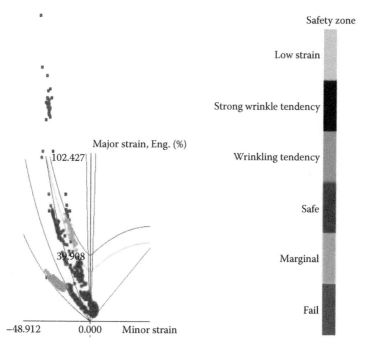

FIGURE 7.11
Tail cap—forming limit diagrams in experiment 9.

7.5 Analysis of Variance

The fracture phenomenon at various cross sections during all nine experiments is presented in Table 7.4. Fracture is calculated from failure limit diagrams. The S/N ratios are calculated for quality characteristics of fracture, where the smaller the ratio, the better is the quality. The addition of the square of vertical distances of all points above the failure-limit curve is made so that all fracture points can be converted to safe points.

The maximum distance of fracture points observed from the failure limit diagrams is 33,108.5 mm^2 and the minimum is 8,144.00 mm^2. The maximum fracture addition of square of distances is observed in experiment 1. The results of the analysis of variance are presented in Tables 7.5 and 7.6. The mean S/N ratios are calculated for all parameters—blank-holder force, coefficient of friction, die profile radius, and punch nose radius—at all three levels, that is, low, medium, and high. The range is defined as the difference between the maximum and minimum values of the S/N ratio for a particular parameter.

Table 7.6 shows a rearrangement of S/N ratios for all variables at all levels. The rank indicates the influence of the input parameter on the output quality characteristic. The result for the orthogonal array indicates that the die profile radius has a major influence on fracture. Friction is second in rank, the blank-holder force is third, and the punch nose radius has the least influence on fracture.

TABLE 7.4

Tail Cap—S/N Ratios in Nine Experiments for Fracture

Experiment No.	Fracture Factor	S/N Ratio
1	33,108.5	−90.39
2	8,144.00	−78.21
3	8,194.34	−78.27
4	8,515.25	−78.60
5	8,615.51	−78.70
6	17,639.11	−84.92
7	10,566.52	−80.47
8	23,172.68	−87.29
9	6,892.82	−76.76

TABLE 7.5

Tail Cap—S/N Ratios at Three Levels for Fracture

Parameter	Level	Experiments	Mean S/N Ratio
Blank-holder force [BHF]	1	1, 2, 3	−82.29
	2	4, 5, 6	−80.74
	3	7, 8, 9	−81.51
Coefficient of friction [μ]	1	1, 4, 7	−83.16
	2	2, 5, 8	−81.40
	3	3, 6, 9	−79.98
Die profile radius [R_D]	1	1, 6, 8	−87.54
	2	2, 4, 9	−77.86
	3	3, 5, 7	−79.15
Punch nose radius [R_P]	1	1, 5, 9	−81.95
	2	2, 6, 7	−81.20
	3	3, 4, 8	−81.39

TABLE 7.6

Tail Cap—ANOVA Results for Fracture

	BHF	Friction	R_D	R_P
1	−82.29	−83.16	−87.54	−81.95
2	−80.74	−81.40	−77.86	−81.20
3	−81.51	−79.98	−79.15	−81.39
Range	1.55	3.18	9.68	0.75
Rank	3	2	1	4

7.6 Objective Function Formulation

Linear mathematical relations have been developed from the results of Taguchi design of experiments and analysis of variance between input parameters like blank-holder force, friction coefficient, die profile radius, and punch nose radius. The performance characteristics applied for the tail cap is fracture. The relationship is presented as follows. Minitab has been used for regression analysis.

The objective function is

Minimize fracture

$$\text{Fracture} = 68{,}925 - \left(294 * \text{BHF}\right) - \left(64{,}880 * \mu\right) - \left(15{,}515 * R_D\right) - \left(1{,}456 * R_P\right)$$

Subject to

$$1.2 \leq \beta \leq 2.2$$

$$3R_D \leq R_P \leq 6R_D$$

$$F_{d\ max} \leq \pi d_m S_0 S_u$$

$$R_D \geq 0.035 \left[50 + \left(d_0 - d_1\right)\right]\sqrt{S_0}$$

7.7 Results

The GA has been applied for optimization. The parameters selected for the algorithm are presented in Table 7.7.

During the minimization process, the diameter of the tail cap and process variables such as die profile radius and coefficient of friction were selected as variables. Table 7.8 presents results of the optimization and optimized value of variables with the lower and upper limits.

TABLE 7.7

Genetic Algorithm Parameters

Population	Double Vector
Selection	Tournament
Crossover	Two point
Mutation	Constraint-dependent
Migration	Forward
Crossover probability	0.80
Pareto front fraction	0.65
Stopping criterion	Number of generations
Generations	700
Initial population	500

TABLE 7.8

Optimization Results—Genetic Algorithm

Parameter	Lower Bound	Upper Bound	Optimum
Friction	0.05	0.15	0.15
Die profile radius [mm]	1.5	2.5	2.5
Diameter of tail cap [mm]	108	112	111
Fracture factor		3235 mm²	

The optimum value of fracture factor achieved is 3235. The optimum diameter obtained is 111 mm. The coefficient friction needs to be maintained at an optimum value of 0.15. The die profile radius obtained is 2.5 mm.

7.8 Validation: Numerical Simulation

The numerical simulation was carried out with the optimum parameters achieved after optimization applying GA. The forming zones observed at different sections in the tail cap are presented in Figure 7.12. The failure limit diagram in Figure 7.13 shows fracture and one may observe that the fracture points have been reduced after optimization.

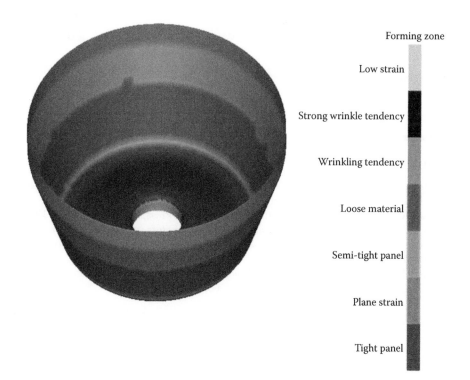

FIGURE 7.12
Tail cap—forming zone in optimized design.

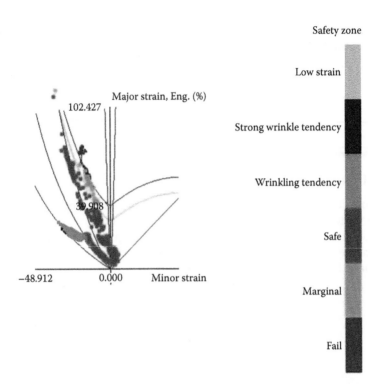

FIGURE 7.13
Tail cap—failure limit diagram for optimized design.

Bibliography

Goldberg, D. E., *Genetic Algorithms in Search, Optimization & Machine Learning*, Pearson Education Pvt. Ltd., Delhi, India, 2003.

Mitchell, M., *An Introduction to Genetic Algorithms*, MIT Press, Cambridge, U.K., 1996.

Spears, W. M. and V. Anand, *A Study of Crossover Operators in Genetic Programming*, Navy Center for Applied Research in AI Naval Research Laboratory, Washington, DC, 2005.

Whitley, D., *Genetic Algorithms and Evolutionary Computing*, Computer Science Department, Colorado State University, Fort Collins, CO.

8

Thinning Optimization of Punch Plate: Gray Wolf Optimizer

8.1 Introduction

The gray wolf optimizer (GWO) is inspired by gray wolves (*Canis lupus*). The GWO algorithm mimics the leadership hierarchy and hunting mechanism of gray wolves in nature. Four types of gray wolves—alpha, beta, delta, and omega—are employed for simulating the leadership hierarchy. In addition, the three main steps of hunting—searching for prey, encircling prey, and attacking prey—are implemented.

8.2 Inspiration

The leaders are a male and a female, called alphas. The alphas are mostly responsible for making decisions about hunting, sleeping place, time to wake, and so on. The alphas' decisions are dictated to the pack. The alpha wolf is also called the dominant wolf since his/her orders should be followed by the pack (Figure 8.1).

The second level in the hierarchy of gray wolves is the beta type. The betas are subordinate wolves that help the alphas in decision-making or other pack activities. The beta wolf can be either male or female, and he/she is probably the best candidate to become an alpha in case one of the alpha wolves dies or becomes very old. The beta wolf respects the alphas, but commands the other lower-level wolves. It plays the role of an adviser to the alpha and discipliner of the pack.

The lowest-ranking gray wolf is the omega type. The omega plays the role of a scapegoat. Omega wolves always have to submit to all the other dominant wolves. In some cases, the omegas are also the babysitters in the pack.

FIGURE 8.1
Hierarchy of gray wolves.

If a wolf is not an alpha, beta, or omega, he/she is called a subordinate or delta. Delta wolves have to submit to alphas and betas, but they dominate the omegas. Scouts, sentinels, elders, hunters, and caretakers belong to this category.

In addition to the social hierarchy of gray wolves, group hunting is another interesting social behavior.

The following are the main phases of gray wolf hunting (Figure 8.2):

- Tracking, chasing, and approaching the prey
- Pursuing, encircling, and harassing the prey until it stops moving
- Attack toward the prey

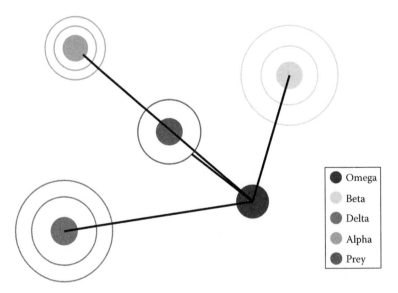

FIGURE 8.2
Position updating in GWO.

8.3 Mathematical Model and Algorithm

8.3.1 Social Hierarchy

In order to mathematically model the social hierarchy of wolves when designing GWO, we consider the fittest solution as alpha (α). Consequently, the second and third best solutions are named beta (β) and delta (δ), respectively. The rest of the candidate solutions are assumed to be omega (ω). In the GWO algorithm, the hunting (optimization) is guided by α, β, δ, and ω.

8.3.2 Encircling Prey

As mentioned, gray wolves encircle their prey during the hunt. In order to mathematically model this encircling behavior, the following equations are proposed:

$$\vec{D} = \left| \vec{C} * \overrightarrow{X_P}(t) - \vec{X}(t) \right| \tag{8.1}$$

$$\vec{X}(t+1) = \overrightarrow{X_P}(t) - \vec{A} * \vec{D} \tag{8.2}$$

where
  t indicates the current iteration
  $\vec{A}$ and $\vec{C}$ are coefficient vectors
  $\overrightarrow{X_P}$ is the position vector of the prey
  $\vec{X}$ indicates the position vector of a gray wolf

The vectors $\vec{A}$ and $\vec{C}$ are calculated as follows:

$$\vec{A} = \left| 2 * \vec{a} * \vec{r_1} - \vec{a} \right| \tag{8.3}$$

$$\vec{C} = 2 * \vec{r_2} \tag{8.4}$$

where components of $\vec{a}$ are linearly decreased from 2 to 0 over the course of iterations and are random vectors in [0, 1].

8.3.3 Hunting

Gray wolves have the ability to recognize the location of prey and encircle them. The hunt is usually guided by the alphas. The betas and deltas might also participate in hunting occasionally. In order to mathematically simulate the hunting behavior of gray wolves, we suppose that the alpha (best candidate solution), beta, and delta have better knowledge about the potential location of prey. Therefore, we save the first three best solutions obtained so far and oblige the other search agents (including the omegas) to update their positions according to the position of the best search agent. The following formulas are proposed in this regard.

$$\vec{D_\alpha} = \left| \vec{C_1} * \vec{X_\alpha} - \vec{X} \right|$$

$$\vec{D_\beta} = \left| \vec{C_2} * \vec{X_\beta} - \vec{X} \right|$$

$$\vec{D_\delta} = \left| \vec{C_3} * \vec{X_\delta} - \vec{X} \right| \tag{8.5}$$

$$\vec{X_1} = \vec{X_\alpha} - \vec{A_1} * \left(\vec{D_\alpha} \right)$$

$$\vec{X_2} = \vec{X_\beta} - \vec{A_2} * \left(\vec{D_\beta} \right)$$

$$\vec{X_3} = \vec{X_\delta} - \vec{A_3} * \left(\vec{D_\delta} \right) \tag{8.6}$$

$$\vec{X}(t+1) = \frac{\vec{X_1} + \vec{X_2} + \vec{X_3}}{3} \tag{8.7}$$

8.3.4 Attacking Prey

The gray wolves finish the hunt by attacking the prey when it stops moving. In order to mathematically model the approach to the prey, we decrease the value of $\bar{a}$. With the operators proposed so far, the GWO algorithm allows its search agents to update their position based on the location of the alpha, beta, and delta and attack the prey.

8.3.5 Search for Prey (Exploration)

Gray wolves mostly search according to the position of the alpha, beta, and delta. They diverge from each other to search for prey and converge to attack prey. In order to mathematically model divergence, we utilize $\vec{A}$ with random values greater than 1 or less than −1 to oblige the search agents to diverge from the prey. This emphasizes exploration and allows the GWO algorithm to search globally (Figure 8.3).

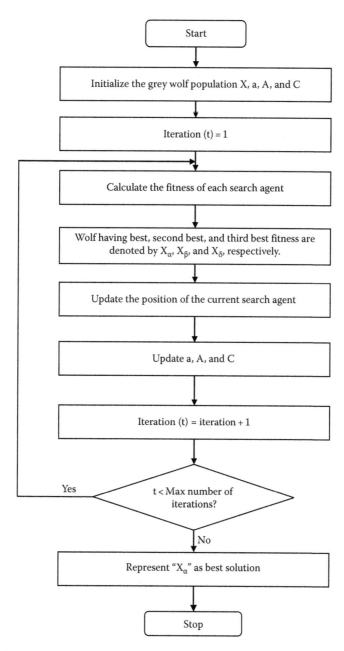

FIGURE 8.3
Flowchart of GWO algorithm.

8.4 Pseudocode

```
Initialize the grey wolf population Xi (i = 1, 2... n)
Initialize a, A, and C
Calculate the fitness of each search agent
    Xα = the best search agent
    Xβ = the second best search agent
    Xδ = the third best search agent
while (t < Max number of iterations)
for each search agent
            Update the position of the current search agent
    end for
    Update a, A, and C
    Calculate the fitness of all search agents
    Update Xα, Xβ, and Xδ
    T = t + 1
end while
return Xα
```

8.5 Penalty Function

Based on mathematical programming approaches, where a constraint numerical optimization problem is transformed into an unconstrained numerical optimization problem (pseudo-objective function), optimization algorithms have adopted penalty functions, whose general formula is

$$\varnothing(\vec{x}) = f(\vec{x}) + p(\vec{x})$$

where $\varnothing(\vec{x})$ is the expanded objective function to be optimized, and $p(\vec{x})$ is the penalty value that can be calculated as follows:

$$p(\vec{x}) = \sum_{i=1}^{m} r_i \cdot \max\left(0, g_i(\vec{x})^2\right) + \sum_{j=1}^{p} c_j \cdot \left|h_j(\vec{x})\right|$$

where r_i and c_j are positive constants called penalty factors.

The main motive of using penalty factors is to favor the selection of feasible solutions over infeasible solutions. There are two types of penalty function methods:

1. Interior penalty function or barrier method
2. Exterior penalty function method

8.5.1 Interior Penalty Function or Barrier Method

The interior penalty function method transforms any constrained optimization problem into an unconstrained one. However, the barrier functions prevent the current solution from ever leaving the feasible region. These require that the interior of the feasible sets be nonempty, which is impossible if equality constraints are present. Therefore, they are used with problems having only inequality constraints.

The following are commonly used penalty functions:

$$B(x) = -\sum_{j=1}^{m} \frac{1}{g_j(x)}$$

or

$$B(x) = -\sum_{j=1}^{m} \log(-g_j(x))$$

The auxiliary function now becomes

$$f(x) + \mu B(x)$$

where μ is a small positive number.

8.5.2 Exterior Penalty Function Method

Like the interior penalty function method this method also transforms any constrained problem into an unconstrained one. The constraints are incorporated into the objective by means of a "penalty parameter," which penalizes any constraint violation. The larger the constraint violation, the larger is the objective function that is penalized.

The penalty function $p(x)$ is defined as

$$p(x) = \sum_{j=1}^{m} \left[\max\{0, g_j(x)\} \right]^{\alpha} + \sum_{j=1}^{m} |h_j(x)|^{\alpha}$$

8.5.3 Static Penalty Function Method

A simple method to penalize infeasible solutions is to apply a constant penalty to those solutions that violate feasibility in any way. The penalized objective function would then be the unpenalized objective function plus a penalty (for a minimization problem). A variation is to construct this simple penalty function as a function of the number of constraints violated where there are multiple constraints. The penalty function for a problem with m constraints would then be

$$f_P(X) = f(X) + \sum_{i=1}^{m} C_i \times \delta_i$$

where
$\delta_i = 1$, if constraint i is violated
$\delta_i = 0$, if constraint i is satisfied
$f_p(x)$ is the penalized objective function
$f(x)$ is the unpenalized objective function
C_i is a constant imposed for violation of constraint i

The following is a general formulation of a minimization problem:

$$f_P(X) = f(X) + \sum_{i=1}^{m} P_i \times \left(g_i(x)\right)^2 + \sum_{i=1}^{m} P_i \times \left(h_i(x)\right)^2$$

where
$g(x)$ is the inequality constraints
$h(x)$ is the equality constraints

8.5.4 Dynamic Penalty Function

In this constraint-handling technique, the individuals are evaluated based on the following formula:

$$eval(\bar{x}) = f(\bar{x}) + (C \times t)^a \sum_{j=1}^{m} f_j^\beta(\bar{x})$$

where C, α, and β are constants.

This method initially penalizes infeasible solutions and, as the iteration number increases, it heavily penalizes infeasible solutions. The penalty component in this method changes with each generation number. However, this method is very sensitive to the parameters C, α, β and these parameters need to be properly tuned before this technique is applied. In this technique, the penalty components have a significant effect on the objective function as they constantly increase with the generation number.

8.6 Punch Plate

The punch plate is manufactured by Vishwadeep Enterprises, Chikhali, Pune. It is part of a two-wheeler's muffler. The punch plate configuration is very interesting but difficult to draw (Figure 8.4). Its circular top has a diameter

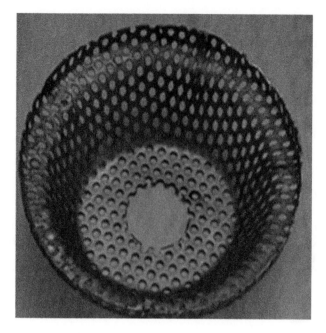

FIGURE 8.4
Punch plate.

of 107.6 mm while its bottom is elliptical, that is, it has a trapezoidal section with a major diameter of 90 mm and minor diameter of 66 mm. The diameter decreases with height. It has a step at the top which is 6 mm high. The corner radius at the junction of the vertical wall and flat portion is 2.5 mm, whereas at the flat portion and inclined wall it is 5 mm. The corner radius at the bottom is also 5 mm. The total height of the cup is 29 mm. The base is flat with a 58 mm diameter. The punch plate doesn't have a flange portion. Table 8.1 presents the details of a punch plate.

TABLE 8.1

Punch Plate Details

Manufactured by	Vishwadeep Enterprises, Pune
Component of	Kinetic Motors Company Ltd.
Part No.	IKS 0025
Weight	50 g
Material	SPCC
Thickness	0.8 mm
Yield strength	280 MPa
Ultimate tensile strength	340 MPa
R	1.3 min
N	0.18 min

8.7 Selection of Process Parameters and Performance Measure: Thinning

Based on a literature survey and industrial insights, four major process parameters have been selected for investigation:

1. Blank-holder force
2. Coefficient of friction
3. Punch nose radius
4. Die profile radius

For the component under study, thinning was selected as the performance measure. Experiments were designed to study and correlate the effect of process parameters on thinning.

8.8 Numerical Investigations: Taguchi Design of Experiments

Nine experiments designed as per L9 orthogonal array [04] have been conducted using finite element simulation software, and the results achieved after experimentation have been used for analyzing the performance parameters. The following are the various types of results achieved for thickness:

- Thickness
- Thickness strain (engineering %)

Every process variable has three levels of operation: low, medium, and high. The orthogonal array selected for this combination of four parameters and three levels is L9. Table 8.2 shows the parameters and their three levels. Table 8.3 presents the detailed L9 orthogonal array.

TABLE 8.2

Punch Plate—Three Levels of Process Parameters

	Lower	Middle	Higher
BHF [kN]	20	25	30
μ	0.05	0.10	0.15
R_D [mm]	4.0	5.0	6.0
R_P [mm]	10.0	12.0	14.0

TABLE 8.3

Punch Plate—L9 Orthogonal Array

Experiment No.	Blank-Holder Force [kN]	Coefficient of Friction	Die Profile Radius [mm]	Punch Nose Radius [mm]
1	20	0.05	4.0	10.0
2	20	0.10	5.0	12.0
3	20	0.15	6.0	14.0
4	25	0.05	5.0	14.0
5	25	0.10	6.0	10.0
6	25	0.15	4.0	12.0
7	30	0.05	6.0	12.0
8	30	0.10	4.0	14.0
9	30	0.15	5.0	10.0

TABLE 8.4

Punch Plate Thickness

Experiment 1	0.79–1.13 mm	Experiment 2	0.76–1.13 mm	Experiment 3	0.73–1.14 mm
Experiment 4	0.78–1.14 mm	Experiment 5	0.77–1.13 mm	Experiment 6	0.73–1.11 mm
Experiment 7	0.79–1.10 mm	Experiment 8	0.77–1.12 mm	Experiment 9	0.72–1.12 mm

The original thickness of the punch plate is 0.8 mm. During experimentation, thinning as well as thickening behavior was observed. The thickness ranges for nine experiments are presented in Table 8.4. Figures 8.5 through 8.13 show the thickness distribution in all nine experiments.

8.9 Analysis of Variance

The decreased thickness at various cross sections during all nine experiments is measured and presented Table 8.5. For analysis of variance the quality characteristic selected is the difference between the original thickness and the decreased thickness. The S/N ratios are calculated for the quality characteristic, where the smaller the ratio, the better is the quality.

The maximum thinning is observed to be 0.762 mm, while the minimum is 0.79 mm. The results of the analysis of variance are presented in Tables 8.6 and 8.7. The mean S/N ratios are calculated for all parameters—blank-holder force, coefficient of friction, die profile radius, and punch nose radius—at all three levels, that is, low, medium, and high. The ANOVA table represents the mean of S/N ratios for all parameters at all three levels from all nine experiments conducted. The range is defined as the difference between the maximum and minimum values of S/N ratios for a particular parameter.

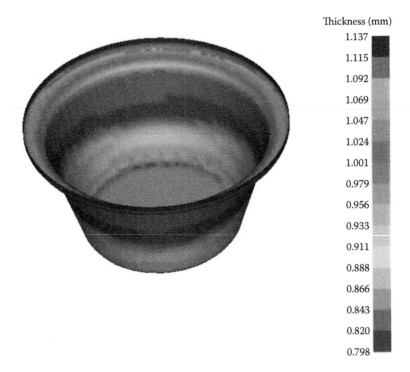

Thickness (mm)
1.137
1.115
1.092
1.069
1.047
1.024
1.001
0.979
0.956
0.933
0.911
0.888
0.866
0.843
0.820
0.798

FIGURE 8.5
Punch plate—thickness distribution in experiment 1.

Thickness (mm)
1.132
1.107
1.083
1.058
1.034
1.009
0.985
0.960
0.936
0.911
0.887
0.862
0.838
0.813
0.789
0.764

FIGURE 8.6
Punch plate—thickness distribution in experiment 2.

Thickness (mm)

1.955
1.874
1.792
1.711
1.630
1.548
1.467
1.386
1.304
1.223
1.141
1.060
0.979
0.897
0.816
0.734

FIGURE 8.7
Punch plate—thickness distribution in experiment 3.

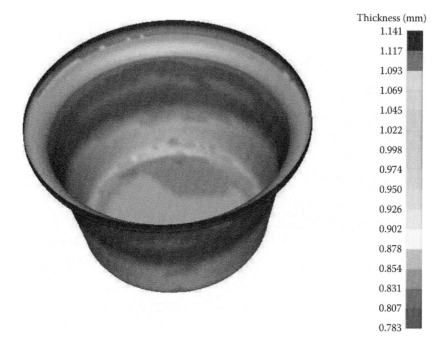

Thickness (mm)

1.141
1.117
1.093
1.069
1.045
1.022
0.998
0.974
0.950
0.926
0.902
0.878
0.854
0.831
0.807
0.783

FIGURE 8.8
Punch plate—thickness distribution in experiment 4.

Thickness (mm)
1.846
1.775
1.704
1.632
1.561
1.489
1.418
1.347
1.275
1.204
1.132
1.061
0.990
0.918
0.847
0.775

FIGURE 8.9
Punch plate—thickness distribution in experiment 5.

Thickness (mm)
1.119
1.093
1.067
1.041
1.015
0.990
0.964
0.938
0.912
0.886
0.860
0.834
0.809
0.783
0.757
0.731

FIGURE 8.10
Punch plate—thickness distribution in experiment 6.

Thickness (mm)
1.716
1.654
1.593
1.532
1.471
1.410
1.349
1.288
1.227
1.166
1.105
1.044
0.982
0.921
0.860
0.799

FIGURE 8.11
Punch plate—thickness distribution in experiment 7.

Thickness (mm)
1.127
1.104
1.081
1.057
1.034
1.010
0.987
0.963
0.940
0.917
0.893
0.870
0.846
0.823
0.800
0.776

FIGURE 8.12
Punch plate—thickness distribution in experiment 8.

Thickness (mm)
1.121
1.094
1.068
1.041
1.015
0.988
0.962
0.935
0.909
0.882
0.856
0.829
0.803
0.776
0.750
0.723

FIGURE 8.13
Punch plate—thickness distribution in experiment 9.

TABLE 8.5

Punch Plate—S/N Ratios in Nine Experiments for Thinning

Decreased Thickness [mm]	Thickness Difference [mm]	S/N Ratio
0.79	0.010	40
0.776	0.024	32.39
0.775	0.025	32.04
0.795	0.005	46.02
0.780	0.020	33.97
0.770	0.030	30.45
0.795	0.005	46.02
0.788	0.012	38.41
0.762	0.038	28.40

Table 8.7 shows rearrangement of S/N ratios for all variables at all levels. The rank indicates the effect of the parameter on the output quality characteristic. The result for this orthogonal array indicates that friction has a major influence on increase in thickness. The punch nose radius holds second rank, the blank-holder force is third, and the die profile radius stands fourth.

TABLE 8.6

Punch Plate—S/N Ratios at Three Levels for Thinning

Parameter	Level	Experiments	Mean S/N Ratio
Blank-holder force [BHF]	1	1, 2, 3	34.81
	2	4, 5, 6	36.81
	3	7, 8, 9	37.61
Coefficient of friction [μ]	1	1, 4, 7	44.02
	2	2, 5, 8	34.92
	3	3, 6, 9	30.29
Die profile radius [R_D]	1	1, 6, 8	36.28
	2	2, 4, 9	35.60
	3	3, 5, 7	37.34
Punch nose radius [R_P]	1	1, 5, 9	34.12
	2	2, 6, 7	36.28
	3	3, 4, 8	38.82

TABLE 8.7

Punch Plate—ANOVA Results for Thinning

	BHF	Friction	R_D	R_P
1	34.81	44.02	36.28	34.12
2	36.81	34.92	35.60	36.28
3	37.61	30.29	37.34	38.82
Range	2.8	13.73	1.74	4.7
Rank	3	1	4	2

8.10 Objective Function Formulation

Linear mathematical relations have been developed from the results of Taguchi design of experiments and analysis of variance between input parameters like blank-holder force, friction coefficient, die profile radius, and punch nose radius. The performance characteristics applied for punch plate is thinning. The relationships are presented as follows. Minitab has been used for regression analysis.

The objective function is

Minimize thinning

$$\text{Thinning} = 0.795 - (0.000133 * \text{BHF}) + (0.243 * \mu) - (0.00033 * R_D) - (0.00217 * R_P)$$

Subject to

$$1.2 \leq \beta \leq 2.2$$
$$3R_D \leq R_P \leq 6R_D$$
$$F_{d\max} \leq \pi d_m S_0 S_u$$
$$R_D \geq 0.035 \left[50 + (d_0 - d_1) \right] \sqrt{S_0}$$

8.11 Results

The GWO was applied for optimization of thinning (Figure 8.14). The parameters selected for the algorithm are presented in Table 8.8.

During the minimization process, the diameter of the punch plate and process variables such as die profile radius and coefficient of friction were selected. Table 8.9 presents the results of optimization and optimized variables along with the lower and upper limits.

The optimum value of thickness achieved is 0.75 mm. So thinning happens up to extent of 0.02 mm. The optimum diameter obtained is 105 mm. The coefficient friction needs to be maintained at an optimum value of 0.14. The die profile radius obtained is 4 mm.

FIGURE 8.14
Sealing cover—optimized geometry.

TABLE 8.8

Gray Wolf Parameters

Cohort Parameters	Set Value
Number of search agents	25
No. of iterations	2000

TABLE 8.9

Optimization Results—Gray Wolf Optimizer

Parameter	Lower Bound	Upper Bound	Optimum
Punch plate diameter [mm]	105	109	105
Die profile radius [mm]	4	6	4
Coefficient of friction	0.005	0.15	0.14
Optimized thickness			0.78 mm

8.12 Validation: Numerical Simulation

The numerical simulation was carried out with the optimum parameters achieved after optimization applying the GWO. The thickness observed at different sections in the punch plate is presented in Figures 8.15 through 8.17. The average thickness in all sections is measured and observed to be 0.78 mm. This indicates that if optimum parameters are selected, the result is lesser thinning.

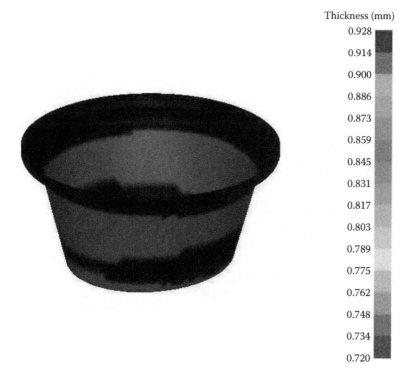

Thickness (mm)

| 0.928 |
| 0.914 |
| 0.900 |
| 0.886 |
| 0.873 |
| 0.859 |
| 0.845 |
| 0.831 |
| 0.817 |
| 0.803 |
| 0.789 |
| 0.775 |
| 0.762 |
| 0.748 |
| 0.734 |
| 0.720 |

FIGURE 8.15
Sealing cover—thickness ranges in optimized design.

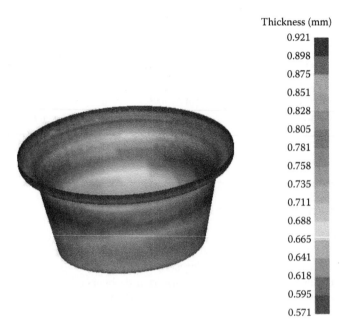

FIGURE 8.16
Sealing cover—thickness ranges in optimized design.

FIGURE 8.17
Sealing cover—thickness ranges in optimized design.

Bibliography

Anusuya Devi, J. and K. Niramathy, Optimal location and sizing of multi type facts devices using grey wolf optimization technique, *International Journal for Scientific Research and Development* 3(03) (2015), 7–9.

EI Gaafary, A. A. M., Y. S. Mohamed, A. M. Hemeida, and A. A. Mohamed, Grey wolf optimization for multi input multi output system, *Universal Journal of Communications and Network* 3(1) (2015), 1–6.

Komaki, G. M. and V. Kayvanfar, Grey wolf optimizer algorithm for the two-stage assembly flow shop scheduling problem with release time, *Journal of Computational Science* 8 (2015), 109–120.

Madadi, A. and M. M. Motlagh, Optimal control of DC motor using grey wolf optimizer algorithm, *Technical Journal of Engineering & Applied Science* 4(4) (2014), 373–379.

Mirjali, S., S. M. Mirjalili, and A. Lewis, Grey wolf optimizer, *Advances in Engineering Software* 69 (2014), 46–61.

Rathee, P., R. Garg, and S. Meena, Using grey wolf optimizer for image registration, *International Journal of Advance Research in Science and Engineering* 4(2, Special Issue) (February 2015), 360–364.

Sharma, S., S. Mehta, and N. Chopra, Economic load dispatch using grey wolf optimization, *International Journal of Engineering Research and Applications* 5(4, Part 6) (April 2015a), 128–132.

Sharma, S., S. Mehta, and N. Chopra, Grey wolf optimization for solving non-convex economic load dispatch, *International Journal of Engineering Research* 3(3) (2015b), 18–21.

Songn, X., L. Tang, S. Zhao, X. Zhang, L. Li, J. Huang, and W. Cai, Grey wolf optimizer for parameter estimation in surface waves, *Soil Dynamics and Earthquake Engineering* 75 (2015), 147–157.

Sulaiman, M. H., Z. Mustaffa, M. R. Mohamed, and O. Aliman, Using the gray wolf optimizer for solving optimal reactive power dispatch problem, *Applied Soft Computing* 32 (2015), 286–292.

Yusof, Y. and Z. Mustaffa, Time series forecasting of energy commodity using grey wolf optimizer, *Proceedings of the International Multi Conference of Engineers and Computer Scientists*, Hong Kong, China, *2015 Vol. I, IMECS 2015*, March 18–20, 2015.

9

Wrinkling Optimization: Firefly Algorithm

9.1 Behavior of Fireflies

The blinking light of fireflies is an incredible sight in the summer sky in the tropical and temperate regions. There exist around 2000 firefly species, and almost all of them produce short and rhythmic flashes. The arrangement of flashes is often unique for a specific species. The flashing light is produced by a process of bioluminescence, and the functions of such signaling systems are still being debated. However, two essential functions of such flashes are to appeal mating partners and to attract possible prey. In addition, flashing may also serve as a defensive cautioning mechanism. The rhythmic flash, the amount of flashing, and the length of period form part of the signaling system that attracts both sexes to each other. Females react to a male's unique pattern of flashing generally within the same species, but in some species, such as photuris, female fireflies can simulate the mating flashing pattern of other species so as to bait and eat the male fireflies who may mistake the flashes as a possible suitable mate. We know that the light intensity at a specific distance r from the light source follows the inverse square law. That is to say, the light intensity I decreases as the distance r increases in terms of $I \propto 1/r^2$. Furthermore, the air absorbs light, which becomes weaker and weaker as the distance increases. These two factors together make most fireflies visible only till a limited distance, usually some hundred meters in the dark, which is usually sufficient for fireflies to communicate. The flashing light can be expressed in such a way that it is associated with the objective function to be optimized, which makes it possible to express new optimization algorithms.

The firefly algorithm (FA) is one of the evolutionary optimization algorithms, and is encouraged by fireflies' behavior in Mother Nature. The FA was developed by Xin-She Yang and it is based on idealized behavior of the flashing forms of fireflies. For simplicity, we can summarize these flashing characteristics into the three following observations:

1. All fireflies are unisex, so that one firefly is attracted to other fireflies regardless of their sex.

2. Attractiveness is proportional to their brightness; thus, for any two flashing fireflies, the less bright one will move toward the brighter one.

3. The attractiveness is proportional to the brightness and they both decrease as their distance increases. If no other firefly is brighter than a particular one, it will move randomly. The brightness of a firefly is affected or determined by the landscape of the objective function to be optimized.

9.2 Firefly Algorithm

Now we can idealize some of the flashing characteristics of fireflies so as to develop firefly-inspired algorithms by using the three observations made earlier. For a maximization problem, the brightness can simply be proportional to the value of the objective function. Other forms of brightness can be defined in a similar way to the fitness function in genetic algorithms. In a certain sense, there is some conceptual similarity between the FA and the bacterial foraging algorithm (BFA). In BFA, the attraction among bacteria is based partly on their fitness and partly on their distance, while in FA, the attractiveness is linked to their objective function and monotonic decay of the attractiveness with distance. However, the agents in FA have adjustable visibility and are more versatile in attractiveness variations, which usually lead to higher mobility, and thus the search space is explored more efficiently.

9.2.1 Attractiveness

In the FA, there are two important issues: the variation of light intensity and the formulation of the attractiveness. For simplicity, we can assume that the attractiveness of a firefly is determined by its brightness, which in turn is associated with the encoded objective function. In the simplest case for maximum optimization problems, the brightness I of a firefly at a particular location x can be considered as $I(x) \propto f(x)$. However, the attractiveness β is relative; it depends on the eyes of the beholder or judged by other fireflies. Thus, it will vary with the distance r_{ij} between firefly i and firefly j. In addition, light intensity decreases with the distance from its source, and light is also absorbed in the media, so we should allow the attractiveness to vary with the degree of absorption. In the simplest form, the light intensity $I(r)$ varies according to the inverse square law $I(r) = Is/r^2$, where Is is the intensity at the source. For a given medium with a fixed light absorption coefficient γ, the light intensity I varies with the distance r. That is $I = I_o e^{-\gamma r}$, where I_o is the original light intensity. In order to avoid the singularity at $r = 0$ in the expression Is/r^2, the combined effect of both the inverse square law and absorption can be approximated using the following Gaussian form:

$$I(r) = I_o e^{-\gamma r^2}$$

Sometimes, we may need a function that decreases monotonically at a slower rate. In this case, we can use the following approximation.

9.2.2 Distance and Movement

The distance between any two fireflies i and j at xi and xj, respectively, is the Cartesian distance:

$$r_{ij} = |X_i - X_j| = \sqrt{\sum_{k=1}^{d} (x_{i,j} - x_{j,k})^2}$$

9.2.3 Scaling and Asymptotic Cases

It is worth pointing out that the distance r defined earlier is not limited to the Euclidean distance. We can define many other forms of distance r in the n-dimensional hyperspace, depending on the type of problem that is of interest to us. For example, for job-scheduling problems, r can be defined as the time lag or time interval. For complicated networks such as the Internet and social networks, the distance r can be defined as the combination of the degree of local clustering and the average proximity of vertices. In fact, any measure that can effectively characterize the quantities of interest in the optimization problem can be used as the "distance" r. The typical scale Γ should be associated with the scale in the optimization problem of interest. If Γ is the typical scale for a given optimization problem, for a very large number of fireflies n > m where m is the number of local optima, the initial locations of these n fireflies should be distributed relatively uniformly over the entire search space in a similar manner to the initialization of quasi–Monte Carlo simulations. As the iterations proceed, the fireflies will converge into all the local optima (including the global ones) in a stochastic manner. By comparing the best solutions among all these optima, the global optima can easily be achieved. At the moment, we are trying to formally prove that the FA will approach global optima when $n \to \infty$ and $t \gg 1$. In reality, it converges very quickly, typically with less than 50–100 generations, and this will be demonstrated using various standard test functions later in this chapter. There are two important limiting cases, namely when $\gamma \to 0$ and $\gamma \to \infty$. For $\gamma \to 0$, the attractiveness is constant $\beta = \beta 0$ and $\Gamma \to \infty$; this is equivalent to saying that the light intensity does not decrease in an idealized sky. Thus, a flashing firefly can be seen anywhere in the domain. Therefore, a single (usually global) optimum can easily be reached. This corresponds to a special case of particle swarm optimization (PSO) discussed earlier. Subsequently, the efficiency of this special case is the same as that of PSO. On the other hand, the limiting case $\gamma \to \infty$ leads to $\Gamma \to 0$ and $\beta(r) \to \delta(r)$ (the Dirac delta function), which means that the attractiveness is almost zero in the sight of other fireflies or the fireflies are short-sighted. This is equivalent to the case where the fireflies fly randomly in a very foggy region. No other fireflies can be seen, and each firefly roams in a completely random manner. Therefore, this corresponds to the completely random search method. As the FA is usually somewhere between these two extremes, it is possible to adjust the parameter γ and α so that it can outperform both the random search and the PSO. In fact, the FA can find the global optima as well as all the local optima simultaneously in a very effective manner. This advantage will be demonstrated in detail later in the implementation. A further advantage of FA is that different fireflies work almost independently; it is thus particularly suitable for parallel implementation. It is even better than genetic algorithms and PSO because fireflies aggregate more closely around each optimum (without jumping around as in the case of genetic algorithms). The interactions between different subregions are minimal in parallel implementation.

9.3 Pseudocode

```
Objective function f(x), x = (x1, ..., xd)T
Generate initial population of fire flies xi (i = 1, 2, ..., n)
Light intensity Ii at xi is determined by f (xi)
Define light absorption coefficient
while (t < Max Generation)
For i = 1 : n all n fireflies
For j = 1 :i all n fireflies
If (Ij > Ii), Move firefly i towards j in d-dimension; end if
Attractiveness varies with distance r via exp[-r]
Evaluate new solutions and update light intensity
end for j
end for i
Rank the fireflies and find the current best
end while
Post process results and visualization
```

9.4 Connector

Connectors are manufactured by Vishwadeep Enterprises, Chikhali, Pune. This is fitted between the fuel neck and the filler nose. The connector is 18 mm long where the top 0.8 mm has a diameter of 48 mm and the bottom 0.8 mm with a diameter of 31 mm. The connecting portion is 2 mm long with 2.5 mm and 3 mm connecting radii.

A connector is shown in Figure 9.1 and Table 9.1 presents its details.

FIGURE 9.1
Connector.

TABLE 9.1

Connector Details

Manufactured by	Vishwadeep Enterprises, Pune
Component of	Tata Engineering and Locomotive Company Co. Ltd.
Part No.	2779 4710 82 08
Weight	20 g
Material	D-513, SS 4010
Thickness	1.00 mm
Yield strength	280 MPa
Ultimate tensile strength	270–410 MPa
r	1.7 min
n	0.22

9.5 Selection of Process Parameters and Performance Measure: Wrinkling

Based on literature survey and industrial insights, four major process parameters have been selected for investigation:

1. Blank-holder force
2. Coefficient of friction
3. Punch nose radius
4. Die profile radius

For the component under study, wrinkling has been designated as a performance measure. Experiments have been designed to study and correlate the effect of process parameters on wrinkling.

9.6 Numerical Investigations: Taguchi Design of Experiments

Nine experiments designed as per L9 orthogonal array have been conducted using finite element simulation software, and the results obtained have been used for analyzing the performance parameters. Wrinkling has been assessed using failure limit diagram with a novel wrinkling factor proposed here. For analysis of variance, the square of vertical distances of all points from the wrinkling curve has been added so as to convert all points from the wrinkling region to a safe region to minimize wrinkling. The least square method is employed for measuring the distance of the wrinkling points. The wrinkling factor is defined in terms of distances, which represent major strain on the failure limit diagram.

Every process variable has three levels of operation: low, medium, and high. The orthogonal array selected for this combination of four parameters and three levels is L9. Table 9.2 presents these parameters and their three levels and Table 9.3 presents the detailed L9 orthogonal array.

TABLE 9.2

Connector—Three Levels of Process Parameters

	Lower	Middle	Higher
BHF	04 KN	05 KN	06 KN
μ	0.05	0.10	0.15
R_D	2.5 mm	3.0 mm	3.5 mm
R_P	8.0 mm	9.0 mm	10.0 mm

TABLE 9.3

Connector—L9 Orthogonal Array

Experiment No.	Blank-Holder Force [KN]	Coefficient of Friction	Die Profile Radius [mm]	Punch Nose Radius [mm]
1	04	0.05	2.5	8.0
2	04	0.10	3.0	9.0
3	04	0.15	3.5	10.0
4	05	0.05	3.0	10.0
5	05	0.10	3.5	8.0
6	05	0.15	2.5	9.0
7	06	0.05	3.5	9.0
8	06	0.10	2.5	10.0
9	06	0.15	3.0	8.0

The forming limit diagrams have been plotted for all nine experiments for the connector (Figures 9.2 through 9.10). Experiments 3, 6, 8, and 9 show failure points. There are points that show wrinkling tendency. The maximum points are in the safe region as well as in the marginal area. The failure occurs at 34.33% and 37.12% of major strain. The combination of maximum major and minor strain is presented in Table 9.4.

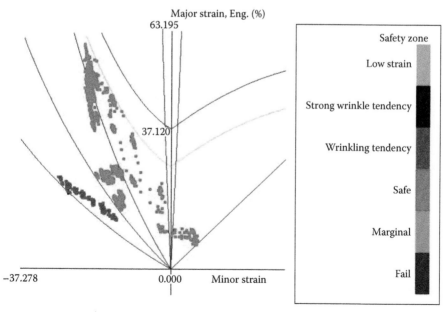

FIGURE 9.2
Connector—forming limit diagrams: first experiment.

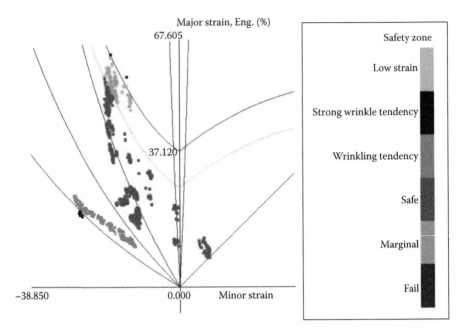

FIGURE 9.3
Connector—forming limit diagrams: second experiment.

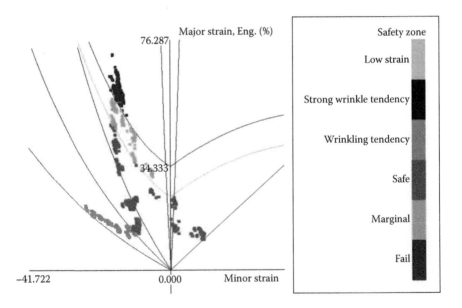

FIGURE 9.4
Connector—forming limit diagrams: third experiment.

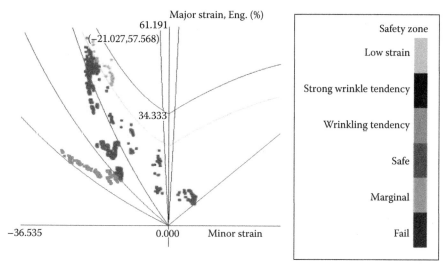

FIGURE 9.5
Connector—forming limit diagrams: fourth experiment.

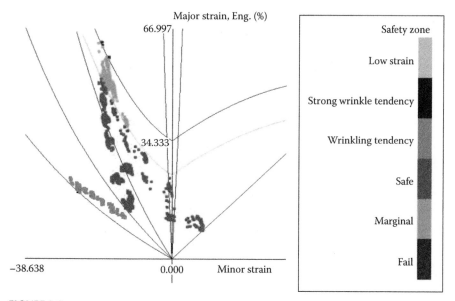

FIGURE 9.6
Connector—forming limit diagrams: fifth experiment.

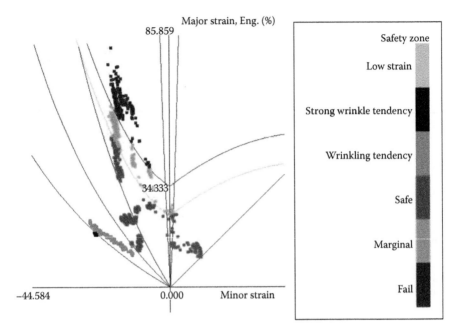

FIGURE 9.7
Connector—forming limit diagrams: sixth experiment.

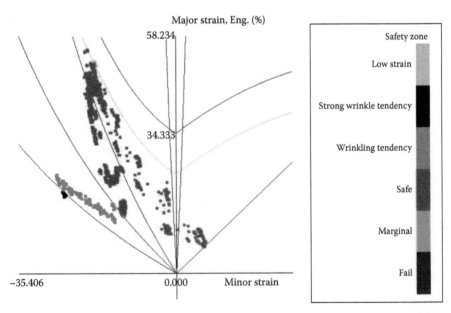

FIGURE 9.8
Connector—forming limit diagrams: seventh experiment.

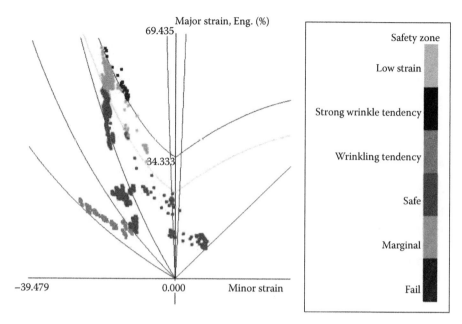

FIGURE 9.9
Connector—forming limit diagrams: eighth experiment.

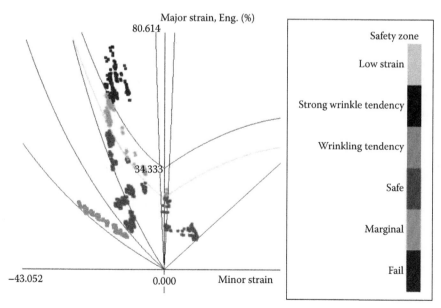

FIGURE 9.10
Connector—forming limit diagrams: ninth experiment.

TABLE 9.4

Connector—Safety Margin Range in Nine Experiments

Experiment 1	−37.2 and 63.1	Experiment 2	−38.8 and 67.6	Experiment 3	−41.7 and 76.2
Experiment 4	−36.5 and 61.1	Experiment 5	−38.6 and 66.9	Experiment 6	−44.5 and 85.8
Experiment 7	−35.4 and 58.2	Experiment 8	−39.4 and 69.4	Experiment 9	−43.0 and 80.6

9.7 Analysis of Variance

The wrinkling factor from all the points from the wrinkling zone in the failure limit diagram during all nine experiments is calculated and presented in Table 9.5. For analysis of variance, the quality characteristic selected is the wrinkling factor. The S/N ratios are calculated for quality characteristics, where the smaller the ratio, the better is the quality.

The maximum distance of the wrinkling points observed from experiment 3 in the failure limit diagram is 1778.80 mm^2 and the minimum is 859.01 mm^2. The results of the analysis of variance are presented Tables 9.6 and 9.7. The mean S/N ratios are calculated for all parameters—blank-holder force, coefficient of friction, die profile radius, and punch nose radius—at low, medium, and high levels. The range is defined as the difference between the maximum and minimum value of the S/N ratio for a particular parameter.

Table 9.7 presents rearrangement of S/N ratios for all variables at all levels. The rank indicates the influence of the input parameter on the quality characteristic. The result for the orthogonal array indicates that the punch nose radius has major influence on wrinkling, the die profile radius ranks second, friction is the third rank, and blank-holder force has the least influence on wrinkling.

TABLE 9.5

Connector—S/N Ratios in Nine Experiments for Wrinkling

Experiment No.	Distance Square of All Wrinkling Points from Wrinkling Curve [mm^2]	S/N Ratio
1	1291.27	−62.22
2	1146.14	−61.18
3	1778.80	−65.00
4	1000.83	−60.00
5	994.69	−59.95
6	859.01	−58.67
7	1307.52	−62.32
8	918.74	−59.26
9	1046.89	−60.39

TABLE 9.6

Connector—S/N Ratios at Three Levels for Wrinkling

Parameter	Level	Experiments	Mean S/N Ratio
Blank-holder force [BHF]	1	1, 2, 3	−63.42
	2	4, 5, 6	−63.88
	3	7, 8, 9	−64.60
Coefficient of friction [μ]	1	1, 4, 7	−64.90
	2	2, 5, 8	−64.00
	3	3, 6, 9	−63.00
Die profile radius [R_D]	1	1, 6, 8	−62.59
	2	2, 4, 9	−64.38
	3	3, 5, 7	−64.93
Punch nose radius [R_P]	1	1, 5, 9	−62.79
	2	2, 6, 7	−63.06
	3	3, 4, 8	−66.05

TABLE 9.7

Connector—ANOVA Results for Wrinkling

	BHF	Friction	R_D	R_P
1	−63.42	−64.90	−62.59	−62.79
2	−63.88	−64.00	−64.38	−63.06
3	−64.60	−63.00	−64.93	−66.05
Range	0.45	1.90	2.34	3.26
Rank	4	3	2	1

9.8 Objective Function Formulation

Linear mathematical relations have been developed from the results of the Taguchi design of experiments and the analysis of variance between the input parameters, namely, blank-holder force, friction coefficient, die profile radius, and punch nose radius. The performance characteristics applied for the connector is wrinkling factor. Minitab has been used for regression analysis and the results are presented.

The objective function is

Minimize wrinkling

$$\text{Wrinkling} = 349 - (157 * \text{BHF}) + (283 * \mu) + (337 * R_D) + (61 * R_P)$$

Subject to

$$1.2 \leq \beta \leq 2.2$$

$$3R_D \leq R_P \leq 6R_D$$

$$F_{d\,max} \leq \pi d_m S_0 S_u$$

$$R_D \geq 0.035 \left[50 + (d_0 - d_1) \right] \sqrt{S_0}$$

9.9 Results

FA has been applied for optimization. The parameters selected for the algorithm are listed in Table 9.8.

During the minimization process, the diameter of the connector and process variables such as die profile radius and coefficient of friction were selected as variables. Table 9.9 presents results of the optimization and the optimized value of the variables with lower and upper limits.

TABLE 9.8

Firefly Algorithm Parameters

Firefly Parameters	Set Value
Firefly population	40
Iterations	500
Light absorption coefficient (γ)	1
Attractiveness (β)	0.2
Randomness scaling factor (α)	0.5

TABLE 9.9

Optimization Results—Flower Pollination

Parameter	Lower Bound	Upper Bound	Optimum
Connector diameter	46 mm	50 mm	47 mm
Corner radius	2.5 mm	3.5 mm	2.55 mm
Coefficient of friction	0.005	0.15	0.02
Wrinkling factor			1125.21

The optimum value of wrinkling factor achieved is 1125. The optimum diameter obtained is 47 mm. The coefficient friction needs to be maintained at the optimum with a value of 0.02. The die profile radius obtained is 2.55 mm.

9.10 Validation: Numerical Simulation

Numerical simulation was conducted with the optimum parameters achieved after optimization with the FA. The forming zones observed at different sections in the connector and safety zones are presented in Figures 9.11 and 9.12. The failure limit diagram in Figure 9.13 shows wrinkling and one may observe that the wrinkling points have reduced after optimization.

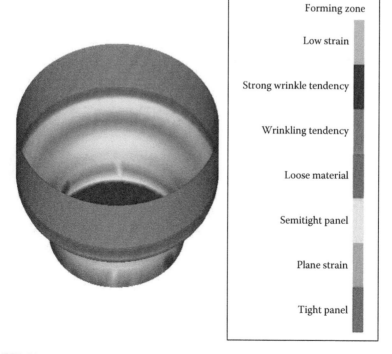

FIGURE 9.11
Optimized connector–forming zone.

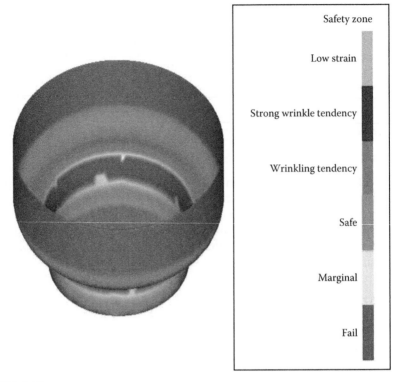

FIGURE 9.12
Optimized connector—safety zone.

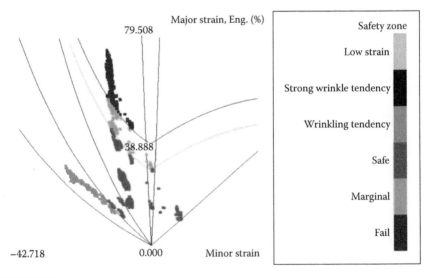

FIGURE 9.13
Optimized connector—failure limit diagram.

Bibliography

Brest, J., I. Fistera, X.-S. Yang, and I. F. Junior, A comprehensive review of firefly algorithms, *Swarm and Evolutionary Computation* 13 (December 2013), 34–46.

Yang, X.S., Firefly algorithms for multimodal optimization, in: *Stochastic Algorithms: Foundations and Applications* (Editors O. Watanabe and T. Zeugmann), SAGA 2009, Lecture Notes in Computer Science, Vol. 5792, Springer, Berlin, Germany, 2009.

Yang, X.-S. and X. He, Firefly algorithm: Recent advances and applications, *International Journal of Swarm Intelligence* 1(1) (2013), 36–50. DOI: 10.1504/IJSI.2013.055801.

Zhang, L., L. Liu, X.-S. Yang, and Y. Dai, A novel hybrid firefly algorithm for global optimization, *PLoS ONE* 11(9) (2016), e0163230. DOI: 10.1371/journal.pone.0163230.

Bibliography

10

Thickness Gradient Optimization: Ant Lion

10.1 Introduction

The ant lion optimization (ALO) algorithm mimics the hunting mechanism of ant lions in nature. Five main steps of hunting prey such as the random walk of ants, building traps, entrapment of ants in traps, catching preys, and rebuilding traps are implemented.

10.1.1 Inspiration

The names of ant lions originate from their unique hunting behavior and their favorite prey. An ant lion larva digs a cone-shaped pit in the sand by moving along a circular path, throwing out the sand with its massive jaw. After digging the trap, the larva hides underneath the bottom of the cone (as a sit-and-wait predator) and waits for insects (preferably ants) to be trapped in the pit. The edge of the cone is sharp enough for insects to fall to the bottom of the trap easily. Once the ant lion realizes that a prey is in the trap, it tries to catch it. When a prey is caught in the jaw, it is pulled under the soil and consumed. After consuming the prey, ant lions throw the leftovers outside the pit and amend the pit for the next hunt. Another interesting behavior that has been observed in the lifestyle of ant lions is that they tend to dig out larger traps as they become hungrier and/or when the moon is full. They have evolved and adapted in this way to improve their chance of survival. The main inspiration of the ALO algorithm comes from the foraging behavior of ant lion larvae.

The ALO algorithm mimics the interaction between ant lions and ants in the trap. To model such interactions, ants are required to move over the search space and ant lions are allowed to hunt them and become fitter using traps. Since ants move stochastically in nature when searching for food, a random walk is chosen for modeling ants' movement as follows:

$$X(\text{iter}) = \left[0, \text{cumsum}\left(2r(1)-1\right), \text{cumsum}\left(2r(2)-1\right), \ldots, \text{cumsum}\left(2r(n)-1\right)\right]$$

$$(10.1)$$

where cumsum calculates the cumulative sum, n is the maximum number of iterations, iter shows the iteration of random walk, and r(t) is a stochastic function defined as follows:

$$r(t)i = 1 \quad \text{if rand} > 0.5$$

$$r(t) = 0 \quad \text{if rand} \leq 0.5 \tag{10.2}$$

The position of ants and their related objective functions are saved in the matrices *MAnt* and *MOA*, respectively. In addition to ants, it is assumed that ant lions also hide somewhere in the search space. In order to save their positions and fitness values, the *MAntlion* and *MOAl* matrices are utilized. The pseudocodes in the ALO algorithm are defined as follows:

Step 1: Initialize the first population of ants and ant lions randomly. Calculate the fitness of ants and ant lions.

Step 2: Find the best ant lion and assume it is the elite. In this study, the best ant lion obtained so far in each iteration is saved and considered to be an elite.

Step 3: For each ant, select an ant lion using the roulette wheel and

    Step 3.1: Create a random walk

    Step 3.2: Normalize them in order to keep the random walks inside the search space

$$X_i^{\text{iter+1}} = \left\{ \left[\left(X_i^{\text{iter}} - a_i \right) \times \left(d_i - a_i \right) \right] \div \left(b_i - a_i \right) \right\} + c_i \tag{10.3}$$

where

    a_i is the minimum random walk of the ith variable

    b_i is the maximum random walk in the ith variable

    c_i and d_i are the minimum and maximum of the ith variable at the current iteration

    Step 3.3: Update the position of ant

$$\text{Ant}_i^{\text{iter+1}} = \left(R_A^{\text{iter}} + R_E^{\text{iter}} \right) \div 2 \tag{10.4}$$

where

    R_A^{iter} is the random walk around the ant lion selected by the roulette wheel

    R_E^{iter} is the random walk around the elite

    $\text{Ant}_i^{\text{iter+1}}$ indicates the position of the ith ant at the iteration iter + 1

Step 3.4: Update c and d using the following:

$$C^{iter+1} = C^{iter} \div I \qquad (10.5)$$

$$d^{iter+1} = d^{iter} \div I \qquad (10.6)$$

where

$$I = 10^W \, iter/n \qquad (10.7)$$

and w is a constant defined based on the current iteration (w = 2 when iter > 0.1n, w = 3 when iter > 0.5n, w = 4 when iter > 0.75n, w = 5 when iter > 0.9n, and w = 6 when iter > 0.95n).

Step 4: Calculate the fitness of all ants.

Step 5: Replace an ant lion with its corresponding ant it if becomes fitter.

Step 6: Update elite if an ant lion becomes fitter than the elite.

Step 7: Repeat from Step 3 until a stopping criterion is satisfied.

10.1.2 Random Walks of Ants

Ants update their positions with random walk at every step of optimization. Since every search space has a boundary (range of variable), equations can't be directly used for updating positions of ants. In order to keep the random walks inside the search space, they are normalized using the following equation (min–max normalization):

$$X_i^t = \frac{\left(X_i^t - a_i\right) \times \left(d_i^t - c_i^t\right)}{\left(b_i - a_i\right)} + c_i^t \qquad (10.8)$$

where
  a_i is the minimum random walk of the ith variable
  b_i is the maximum random walk in the ith variable
  c_i^t is the minimum of the ith variable at the tth iteration
  d_i^t indicates the maximum of the ith variable at the tth iteration

10.1.3 Trapping in Ant Lions' Pits

Random walks of ants are affected by ant lions' traps. In order to mathematically model this assumption, the following equations are proposed:

$$c_i^t = Ant\ lion_j^t + c^t \qquad (10.9)$$

$$d_i^t = \text{Ant lion}_j^t + d^t \qquad\qquad (10.10)$$

where

c^t is the minimum of all variables at the tth iteration

d^t indicates the vector including the maximum of all variables at the tth iteration

c_i^t is the minimum of all variables for the ith ant

d_i^t is the maximum of all variables for the ith ant

Ant lion$_j^t$ shows the position of the selected jth ant lion at the tth iteration

Equations 10.9 and 10.10 show that ants randomly walk in a hyper-sphere defined by the vectors c and d around a selected ant lion.

10.1.4 Building Trap

In order to model the ant lion's hunting capability, a roulette wheel is employed. Ants are assumed to be trapped in only one selected ant lion. The ALO algorithm is required to utilize a roulette wheel operator for selecting ant lions based on their fitness during optimization. This mechanism gives high chances to the fitter ant lions for catching ants.

10.1.5 Sliding Ants toward Ant Lion

With the mechanisms proposed so far, ant lions are able to build traps proportional to their fitness and ants are required to move randomly. However, ant lions shoot sand outward from the center of the pit once they realize that an ant is in the trap. This behavior slides down the trapped ant that is trying to escape. For mathematically modeling this behavior, the radius of an ant's random walk hyper-sphere is decreased adaptively. The following equations are proposed in this regard:

$$c^t = \frac{c^t}{I} \qquad\qquad (10.11)$$

$$d^t = \frac{d^t}{I} \qquad\qquad (10.12)$$

10.1.6 Catching Prey and Rebuilding the Pit

The final stage of the hunt is when an ant reaches the bottom of the pit and is caught in the ant lion's jaw. After this stage, the ant lion pulls the ant inside the sand and consumes its body. For mimicking this process, it is assumed that catching prey occurs when an ant becomes fitter (goes inside sand) than its corresponding ant lion. An ant lion is then required to update its position to the latest position of the hunted ant to enhance its chance of catching new prey. The following equation is proposed in this regard:

$$\text{Ant lion}_j^t = \text{Ant}_i^t \quad \text{if } f\left(\text{Ant}_i^t\right) > f\left(\text{Ant lion}_j^t\right) \tag{10.13}$$

where
t shows the current iteration
Ant lion$_j^t$ shows the position of the selected jth antlion at the tth iteration
Ant$_i^t$ indicates the position of the ith ant at the tth iteration

10.1.7 Elitism

Elitism is an important characteristic of evolutionary algorithms that allows them to maintain the best solution obtained at any stage of the optimization process. In this study, the best ant lion obtained so far in each iteration is saved and considered as an elite. Since the elite is the fittest ant lion, it should be able to affect the movements of all the ants during iterations. Therefore, it is assumed that every ant randomly walks around an ant lion selected by the roulette wheel and the elite simultaneously, as follows:

$$\text{Ant}_i^t = \frac{R_A^t + R_E^t}{2} \tag{10.14}$$

where
R_A^t is the random walk around the ant lion selected by the roulette wheel at the tth iteration
R_E^t is the random walk around the elite at the tth iteration
Ant$_i^t$ indicates the position of the ith ant at the tth iteration

10.2 Pseudocode of ALO Algorithm

```
1) Initialize the first population of ants and ant lions
   randomly.
2) Calculate the fitness of ants and ant lions.
3) Find the best ant lions and assume it as the elite
   (determined optimum)
4) while the end criterion is not satisfied
   for every ant
   Select an ant lion using Roulette wheel
   Update c and d using Equations 10.11 and 10.12
   Create a random walk and normalize it using
   Equations 10.1 and 10.8
   Update the position of ant using Equations 10.14.
```

```
            end for
            Calculate the fitness of all ants
            Replace an ant lion with its corresponding ant if it
            becomes
            fitter (Equation 10.13)
        5)  Update elite if an ant lion becomes fitter than the
            elite
            end while
            Return elite
```

10.3 Tail Cap

The tail cap is manufactured by Vishwadeep Enterprises, Chikhali, Pune. It is fitted at the exhaust end of a two-wheeler's silencer. The tail cap configuration is most complicated (Figure 10.1). It has a trapezoidal section with a top diameter of 110 mm and its bottom is elliptical with a major diameter of 80 mm and a minor diameter of 70 mm. The total height of the cup is 43 mm, where the upper 8 mm is the straight portion while the remaining is inclined. The corner radius between the inclined wall and the straight wall is 2.7 mm. The bottom is also inclined with a central hole of 21.5 mm diameter emerging outward. The corner radius at the bottom is 5 mm and in the radius between the inclined bottom and the hole is 2 mm. The hole has a height of 5.8 mm.

Table 10.1 provides details of the tail cap.

FIGURE 10.1
Tail cap.

TABLE 10.1

Tail Cap Details

Manufactured by	Vishwadeep Enterprises, Pune
Component of	Kinetic Motors Company Ltd.
Part No.	IKS 0024A
Weight	166 g
Material	SPCD
Thickness	1.2 mm
Yield strength	250 MPa
Ultimate tensile strength	270–370 MPa
r	1.3 min
n	0.18 min

10.4 Selection of Process Parameters and Performance Measure: Thickness Gradient

Based on a literature survey and industrial insights, four major process parameters have been selected for investigation:

1. Blank-holder force
2. Coefficient of friction
3. Punch nose radius
4. Die profile radius

Thickness gradient has been selected as the performance measure. Experiments are designed to study and correlate the effects of process parameters on uniform thickness distribution.

10.5 Numerical Investigations: Taguchi Design of Experiments

Nine experiments designed as per L9 orthogonal array have been conducted using finite element simulation software, and the results achieved after experimentation have been used for analyzing the performance parameters (Figures 10.2 through 10.10). Thickness gradient has been defined as the ratio of thickness at two adjacent nodes in the finite element analysis approach or two circles in practical circle grid analysis.

Every process variable has three levels of operation: low, medium, and high. The orthogonal array selected for this combination of four parameters and three levels is L9. Table 10.2 shows the parameters and their three levels. Table 10.3 presents the detailed L9 orthogonal array.

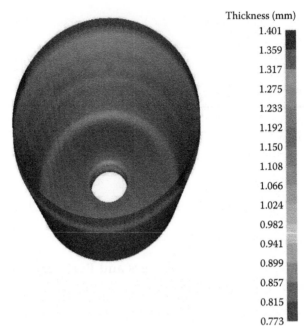

FIGURE 10.2
Tail cap—thickness distribution in experiment 1.

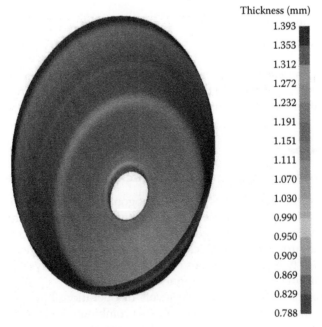

FIGURE 10.3
Tail Cap—thickness distribution in experiment 2.

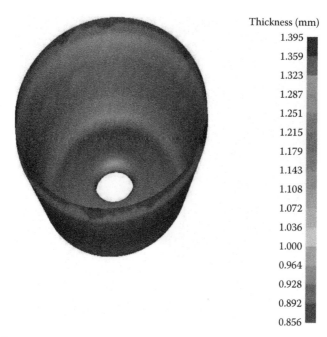

Thickness (mm)
1.395
1.359
1.323
1.287
1.251
1.215
1.179
1.143
1.108
1.072
1.036
1.000
0.964
0.928
0.892
0.856

FIGURE 10.4
Tail Cap—thickness distribution in experiment 3.

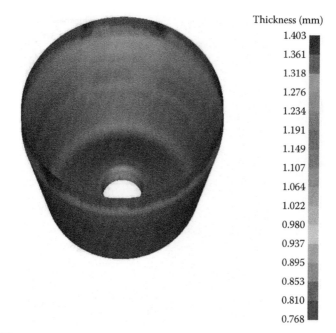

Thickness (mm)
1.403
1.361
1.318
1.276
1.234
1.191
1.149
1.107
1.064
1.022
0.980
0.937
0.895
0.853
0.810
0.768

FIGURE 10.5
Tail cap—thickness distribution in experiment 4.

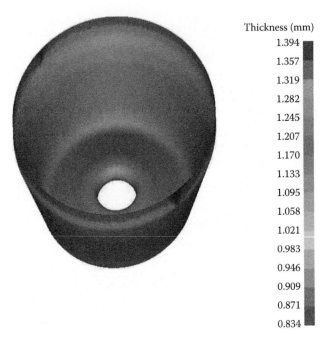

Thickness (mm)
1.394
1.357
1.319
1.282
1.245
1.207
1.170
1.133
1.095
1.058
1.021
0.983
0.946
0.909
0.871
0.834

FIGURE 10.6
Tail cap—thickness distribution in experiment 5.

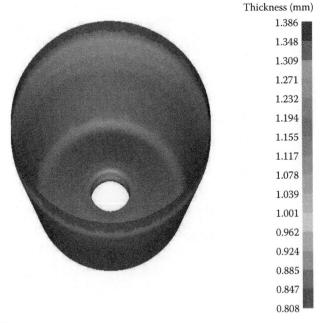

Thickness (mm)
1.386
1.348
1.309
1.271
1.232
1.194
1.155
1.117
1.078
1.039
1.001
0.962
0.924
0.885
0.847
0.808

FIGURE 10.7
Tail cap—thickness distribution in experiment 6.

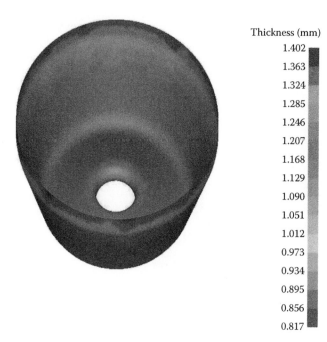

Thickness (mm)

1.402
1.363
1.324
1.285
1.246
1.207
1.168
1.129
1.090
1.051
1.012
0.973
0.934
0.895
0.856
0.817

FIGURE 10.8
Tail cap—thickness distribution in experiment 7.

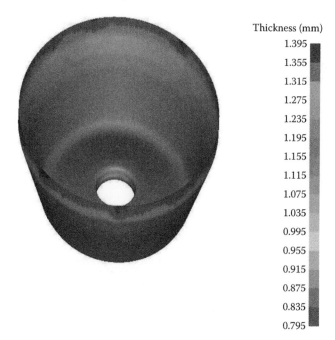

Thickness (mm)

1.395
1.355
1.315
1.275
1.235
1.195
1.155
1.115
1.075
1.035
0.995
0.955
0.915
0.875
0.835
0.795

FIGURE 10.9
Tail cap—thickness distribution in experiment 8.

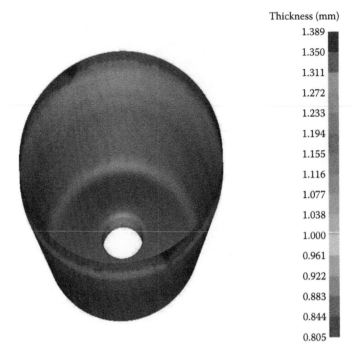

Thickness (mm)

| 1.389 |
| 1.350 |
| 1.311 |
| 1.272 |
| 1.233 |
| 1.194 |
| 1.155 |
| 1.116 |
| 1.077 |
| 1.038 |
| 1.000 |
| 0.961 |
| 0.922 |
| 0.883 |
| 0.844 |
| 0.805 |

FIGURE 10.10
Tail cap—thickness distribution in experiment 9.

The original thickness of the tail cap is 1.0 mm. During experimentation thinning as well as thickening behavior was observed. The thickness ranges for the nine experiments are presented in Table 10.4.

The thickness variation can be seen from 0.76 to 1.40 mm in various regions of the tail cap. The thickness at the bottom is 1.13 mm and increases to 1.17 mm in the bottom region of the wall. The thickening phenomenon is also observed in the upper portion of the wall, and the thickness at the top of the wall is 1.33 mm.

TABLE 10.2

Tail Cap—Three Levels of Process Parameters

	Lower	Middle	Higher
BHF [kN]	20	25	30
μ	0.05	0.10	0.15
R_D [mm]	1.5	2.0	2.5
R_P [mm]	6.0	7.0	8.0

TABLE 10.3

Tail Cap—L9 Orthogonal Array

Experiment No.	Blank-Holder Force [kN]	Coefficient of Friction	Die Profile Radius [mm]	Punch Nose Radius [mm]
1	20	0.05	1.5	6.0
2	20	0.10	2.0	7.0
3	20	0.15	2.5	8.0
4	25	0.05	2.0	8.0
5	25	0.10	2.5	6.0
6	25	0.15	1.5	7.0
7	30	0.05	2.5	7.0
8	30	0.10	1.5	8.0
9	36	0.15	2.0	6.0

TABLE 10.4

Tail Cap—Thickness Distribution

Experiment 1	0.77–1.40 mm	Experiment 2	0.78–1.39 mm	Experiment 3	0.85–1.39 mm
Experiment 4	0.76–1.40 mm	Experiment 5	0.83–1.39 mm	Experiment 6	0.80–1.38 mm
Experiment 7	0.81–1.40 mm	Experiment 8	0.79–1.39 mm	Experiment 9	0.80–1.38 mm

10.6 Analysis of Variance

For calculation of thickness gradients, a cross section of the component has been taken and thickness gradients have been calculated by measuring actual thickness during numerical experimentation. The average thickness gradient recorded during all nine experiments for variability analysis of the process is presented in Table 10.5. The mean thickness gradient for each experiment

TABLE 10.5

Tail Cap—S/N Ratios in Nine Experiments for Thickness Gradient

Experiment No.	Thickness Gradient Mean	S/N Ratios
1	0.979722	33.03
2	0.987827	38.23
3	0.982265	37.24
4	0.993044	46.32
5	0.973410	31.37
6	0.993242	36.68
7	0.997756	59.94
8	0.9796852	50.54
9	0.998932	59.43

is calculated and then S/N ratios are determined. The quality characteristic of thickness gradient is considered for analysis of variance, where a nominal value is considered to be better.

The maximum thickness gradient recorded is 1.0 and the minimum is 0.94. The average maximum thickness gradient of 0.998 is observed in experiment 9. The results of the analysis of variance are presented in Tables 10.6 and 10.7. The mean S/N ratios are calculated for all parameters—blank-holder force, coefficient of friction, die profile radius, and punch nose radius—at all three levels, that is, low, medium, and high. The range is defined as the difference between the maximum and minimum values of mean S/N ratios for a particular parameter.

Table 10.7 shows the rearrangement of S/N ratios for all variables at all levels. The rank indicates the influence of the input parameter on the quality characteristic. The result for the orthogonal array indicates that the blank-holder force has a major influence on the thickness gradient. The die profile radius is second in rank, friction is the third, and the punch nose radius has the least influence on the thickness gradient.

TABLE 10.6

Tail Cap—S/N Ratios at Three Levels for Thickness Gradient

Parameter	Level	Experiments	Mean S/N Ratio
Blank-holder force [BHF]	1	1, 2, 3	36.16
	2	4, 5, 6	38.12
	3	7, 8, 9	56.63
Coefficient of friction [μ]	1	1, 4, 7	46.43
	2	2, 5, 8	40.04
	3	3, 6, 9	44.45
Die profile radius [R_D]	1	1, 6, 8	40.08
	2	2, 4, 9	47.99
	3	3, 5, 7	42.85
Punch nose radius [R_P]	1	1, 5, 9	41.27
	2	2, 6, 7	41.81
	3	3, 4, 8	44.70

TABLE 10.7

Tail Cap—ANOVA Results for Thickness Gradient

	BHF	Friction	R_D	R_P
1	36.16	46.43	40.08	41.27
2	38.12	40.04	47.99	41.81
3	56.63	44.45	42.85	44.70
Range	20.47	6.39	7.91	3.43
Rank	1	3	2	4

10.7 Objective Function Formulation

Linear mathematical relations have been developed from the results of Taguchi design of experiments and analysis of variance between input parameters like blank-holder force, friction coefficient, die profile radius, and punch nose radius. The performance characteristic applied for tail cap is thickness gradient. The relationship is presented as follows. Minitab has been used for regression analysis.

The objective function is

Minimize thickness gradient

$$\text{Thickness gradient} = 0.943 + (0.000667 * \text{BHF}) + (0.0333 * \mu) + (0.00333 * R_D)$$
$$+ (0.00167 * R_P)$$

Subject to

$$1.2 \leq \beta \leq 2.2$$
$$3R_D \leq R_P \leq 6R_D$$
$$F_{d \ max} \leq \pi d_m S_0 S_u$$
$$R_D \geq 0.035 \left[50 + (d_0 - d_1) \right] \sqrt{S_0}$$

10.8 Results

ALO has been applied for optimization. The parameters selected for the algorithm are shown in Table 10.8.

During the minimization process, the diameter of the tail cap and process variables such as die profile radius and coefficient of friction were selected as variables. Table 10.9 presents the results of optimization and optimized value of variables with the lower and upper limits.

The optimum value of thickness gradient achieved is 0.98. The optimum diameter obtained is 112 mm. The coefficient friction needs to be maintained at an optimum value of 0.005. The die profile radius obtained is 3.99 mm.

TABLE 10.8

Ant Lion Parameters

Ant Lion Parameters	Set Value
Number of search agents	50
Number of iterations	500

TABLE 10.9

Optimization Results—Ant Lion Algorithm

Parameter	Lower Bound	Upper Bound	Optimum
Tail cap diameter [mm]	108	112	112
Corner radius [mm]	2	4	3.99
Coefficient of friction	0.005	0.15	0.005
Optimized thickness gradient			0.98

10.9 Validation: Numerical Simulation

The numerical simulation was carried out with the optimum parameters achieved after optimization applying ALO. The thickness distributions observed at different sections in the tail cap are presented in Figures 10.11 through 10.15. Thickness gradients were calculated at two different sections on the components and the mean thickness gradient observed is 0.97. This indicates that with optimized parameters there is uniform thickness distribution across the component.

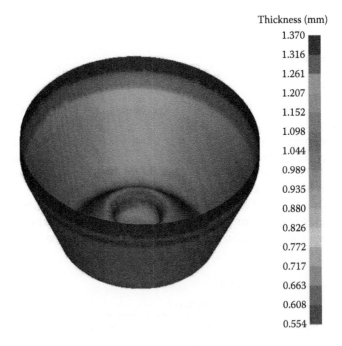

Thickness (mm)
1.370
1.316
1.261
1.207
1.152
1.098
1.044
0.989
0.935
0.880
0.826
0.772
0.717
0.663
0.608
0.554

FIGURE 10.11
Thickness distribution in optimized geometry.

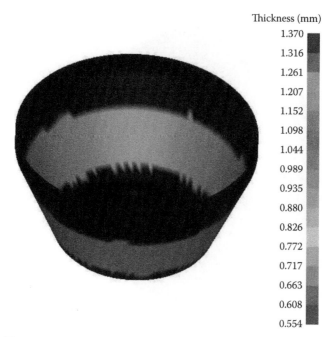

FIGURE 10.12
Thickness distribution ranges in optimized geometry.

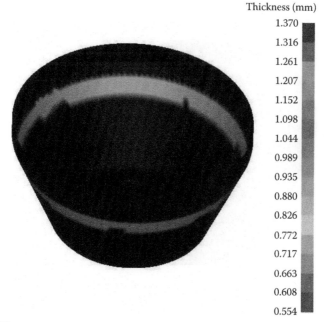

FIGURE 10.13
Thickness distribution ranges in optimized geometry.

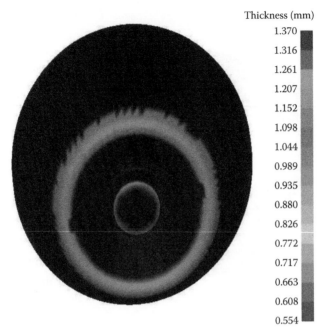

FIGURE 10.14
Thickness distribution ranges in optimized geometry.

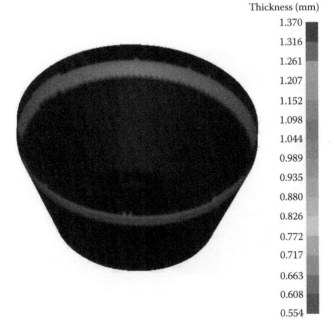

FIGURE 10.15
Thickness distribution ranges in optimized geometry.

Bibliography

Fertin, A. and J. Casas, Orientation towards prey in antlions: Efficient use of wave propagation in sand, *The Journal of Experimental Biology* 210 (July 2007), 3337–3343, DOI: 10.1242/jeb.004473.

Inon, S., S. Aziz, and O. Ofer, Foraging behaviour and habitat selection in pit building ant lion larvae in constant light or dark conditions, *Animal Behaviour* 76(6) (December 2008), 2049–2057.

Mirjalili, S., The ant lion optimizer, *Advances in Engineering Software* 83(C) (March 2015), 80–98, DOI: 10.1016/j.advengsoft.2015.01.010.

Nischal, M. M. and S. Mehta, Optimal load dispatch using ant lion optimization, *International Journal of Engineering Research and Applications* 5(8, Part 2) (August 2015), 10–19.

Talatahari, S., Optimum design of skeletal structures using ant lion optimizer, *International Journal of Optimization in Civil Engineering* 6(1) (2016), 16–25.

11

Springback Optimization: Cuckoo Search

11.1 Cuckoo Breeding Behavior

Cuckoos are charming birds, not only because of the lovely sounds they make, but also because of their aggressive reproduction tactic. Some species such as the Ani and Guira cuckoos lay their eggs in common nests, though they may eradicate other eggs to increase the hatching possibility of their own eggs. To a certain extent, a number of species are involved in obligate brood parasitism by laying their eggs in the nests of other host birds (often other species). There are three basic types of brood parasitism: intraspecific brood parasitism, cooperative breeding, and nest takeover. Some host birds can engage in straight conflict with the intruding cuckoos. If a host bird finds that the eggs are not its own, it will either toss away the unfamiliar eggs or simply undo its nest and build a new nest elsewhere. Some cuckoo species such as the new world brood parasitic Tapera have evolved in such a way that female parasitic cuckoos are often very specialized in the color and pattern imitation of the eggs of a few selected host species. This decreases the probability of their eggs being unrestrained and thus increases their reproductivity. In addition, the timing of egg laying of some species is also astonishing. Parasitic cuckoos frequently choose a nest where the host bird has just laid its own eggs. In general, cuckoo eggs hatch somewhat earlier than their host eggs. Once the first cuckoo chick emerges, the first instinctive action it will take is to expel the host eggs by recklessly propelling the eggs out of the nest, which increases the cuckoo chick's portion of food provided by its host bird. Studies also show that a cuckoo chick can imitate the call of host chicks to gain better access to more feeding.

11.2 Lévy Flights

On the other hand, various studies have shown that the flying behavior of many animals and insects has established the typical characteristics of Lévy flights. A current study by Reynolds and Frye indicates that fruit flies or

Drosophila melanogaster search their landscape by means of series of straight flight trails interrupted by a sudden 90° turn, leading to a Lévy flight–style intermittent scale-free search pattern. Research on human behavior in hunter-gatherer searching patterns also shows the typical feature of Lévy flights. Even light can be related to Lévy flights. Subsequently, such behavior has been applied to optimization and optimal search, and preliminary results show its promising ability.

11.3 Cuckoo Search

For ease of describing the cuckoo search (CS), we use the following three ideal rules:

1. Each cuckoo lays one egg at a time, and dumps its egg in a randomly chosen nest.
2. The best nests with high quality of eggs will carry over to the next generations.
3. The number of available host nests is fixed, and the egg laid by a cuckoo is discovered by the host bird with a probability $P_a \in [0, 1]$.

In this case, the host bird can also throw the egg away or abandon the nest, and build a totally new nest. For simplicity, this last hypothesis can be approximated by the fraction P_a of the n nests that are substituted by new nests (with new random solutions). For a maximization problem, the quality or fitness of a solution can just be proportional to the value of the objective function. Other forms of fitness can be clear in a similar way to the fitness function in genetic algorithms. For simplicity, we can use the following simple representations that each egg in a nest signifies a solution, and a cuckoo egg signifies a new solution; so the aim is to use the new and potentially better solutions (cuckoos) to swap a not-so-good solution in the nests. Of course, this algorithm can be extended to the more intricate case where each nest has multiple eggs signifying a set of solutions. When generating new solutions x(t + 1) for, say, a cuckoo i, a Lévy flight is performed

$$x_i^{(t+1)} = x_i^t + \alpha \oplus \text{Lévy}(\lambda)$$

where $\alpha > 0$ is the step size that should be related to the scales of the problem of interests. In most cases, we can use $\alpha = 1$. The given equation is essentially the stochastic equation for a random walk. In general, a random walk is a Markov chain whose next status/location only depends on the current location (the first term in the equation) and the transition probability (the second term). The product $\oplus$ means entry-wise multiplications. This entry-wise product is similar to those used in particle swarm optimization (PSO), but here the random walk via Lévy flight is more efficient in exploring the search space as its step length is much longer in the long run. The Lévy flight essentially provides a random walk while the random step length is drawn from a Lévy distribution

$$\text{Lévy} \sim u = t^{-\lambda} \qquad (1 < \lambda < 3)$$

which has an infinite variance with an infinite mean. Here the steps essentially form a random walk process with a power law step-length distribution with a heavy tail. Some of the new solutions should be generated by Lévy walk around the best solution obtained so far as this will speed up the local search. However, a substantial fraction of the new solutions should be generated by far-field randomization and whose locations should be far enough from the current best solution, as this will make sure the system will not be trapped in a local optimum. From a quick look, it seems that there is some similarity between CS and hill climbing in combination with some large-scale randomization. But there are some significant differences. Firstly, CS is a population-based algorithm, in a way similar to the genetic algorithm (GA) and PSO, but it uses some sort of elitism and/or selection similar to that used in the harmony search. Secondly, the randomization is more efficient as the step length is heavy-tailed, and any large step is possible. Thirdly, the number of parameters to be tuned is less than with GA and PSO, and thus it is potentially more generic to adapt to a wider class of optimization problems. In addition, each nest can represent a set of solutions; CS can thus be extended to the type of meta-population algorithm. The algorithm is presented as follows.

```
begin
      Objective Function f(x), x = (x₁, x₂………. xd)ᵀ
      Generate initial population of
      n host nests xᵢ = (i = 1, 2, ………n)
while
      (t < Max Generation) or (Stop Criteria)
      Get a cuckoo randomly by Levy flights
      Evaluate its quality/fitness Fᵢ
      Choose a nest among n (say, j) randomly
      If (Fᵢ > Fⱼ).
            Replace j by the new solution;
End
```

```
A fraction (pₐ)of worse nests
     Are abandoned and new ones are built;
Keep the best solutions
     (or nests with quality solutions);
Rank the solutions and find the current best
End while
Post process results and visualization
end
```

11.4 Punch Plate

The punch plate is manufactured by Vishwadeep Enterprises, Chikhali, Pune. It is part of a two-wheeler's muffler. The punch plate configuration is very interesting but difficult to draw (Figure 11.1). Its top has a circular diameter of 107.6 mm, while its bottom is elliptical, that is, it has trapezoidal section with a major diameter of 90 mm and a minor diameter of 66 mm. The diameter decreases with height. It has a step at the top that is 6 mm high. The corner radius at the junction of the vertical wall and the flat portion is 2.5 mm, whereas at the flat portion and the inclined wall it is 5 mm.

FIGURE 11.1
Punch plate.

TABLE 11.1

Punch Plate Details

Manufactured by	Vishwadeep Enterprises, Pune
Component of	Kinetic Motors Company Ltd.
Part No.	IKS 0025
Weight	50 g
Material	SPCC
Thickness	0.8 mm
Yield strength	280 MPa
Ultimate tensile strength	340 MPa
R	1.3 min
N	0.18 min

The corner radius at the bottom is also 5 mm. The total height of the cup is 29 mm. The base is flat with a 58 mm diameter. The punch plate doesn't have a flange portion. Table 11.1 provides details of the punch plate.

11.5 Selection of Process Parameters and Performance Measure: Thinning

Based on a literature survey and industrial insights, four major process parameters have been selected for investigation:

1. Blank-holder force
2. Coefficient of friction
3. Punch nose radius
4. Die profile radius

For the component under study thinning is selected as the performance measure. Experiments are designed to study and correlate the effect of process parameters on thinning.

11.6 Numerical Investigations: Taguchi Design of Experiments

Nine experiments designed as per L9 orthogonal array [04] have been conducted using finite element simulation software, and the results achieved after experimentation have been used for analyzing the performance parameters. Springback displacement magnitude is measured during experimentation.

TABLE 11.2

Punch Plate—Three Levels of Process *Parameters*

	Lower	Middle	Higher
BHF [kN]	20	25	30
μ	0.05	0.10	0.15
R_D [mm]	4.0	5.0	6.0
R_P [mm]	10.0	12.0	14.0

TABLE 11.3

Punch Plate—L9 Orthogonal Array

Experiment No.	Blank-Holder Force [kN]	Coefficient of Friction	Die Profile Radius [mm]	Punch Nose Radius [mm]
1	20	0.05	4.0	10.0
2	20	0.10	5.0	12.0
3	20	0.15	6.0	14.0
4	25	0.05	5.0	14.0
5	25	0.10	6.0	10.0
6	25	0.15	4.0	12.0
7	30	0.05	6.0	12.0
8	30	0.10	4.0	14.0
9	30	0.15	5.0	10.0

Every process variable has three levels of operation: low, medium, and high. The orthogonal array selected for this combination of four parameters and three levels is L9. Table 11.2 presents the parameters and their three levels. Table 11.3 presents the detailed L9 orthogonal array.

The springback displacement magnitude for the punch plate in all nine designed experiments ranges from 0.011 mm to a maximum of 0.21 mm (Figures 11.2 through 11.10). It is observed to be positive in all regions of the component. The range for every experiment is presented in Table 11.4.

In experiment 5, the springback displacement magnitude in the wall region varies in the range of 0.027–0.029, 0.037–0.048, and 0.048–0.58 mm at different locations. In the upper stepped area, it is in the range of 0.11–0.12 and 0.12–0.13 mm. There is a ring at the top portion of the step, where it is very high in the range of 0.16–0.17 mm.

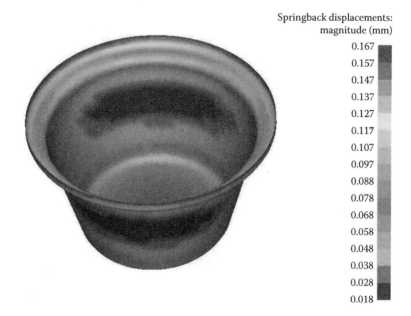

Springback displacements:
magnitude (mm)

0.167
0.157
0.147
0.137
0.127
0.117
0.107
0.097
0.088
0.078
0.068
0.058
0.048
0.038
0.028
0.018

FIGURE 11.2
Punch plate—springback displacement magnitude in experiment 1.

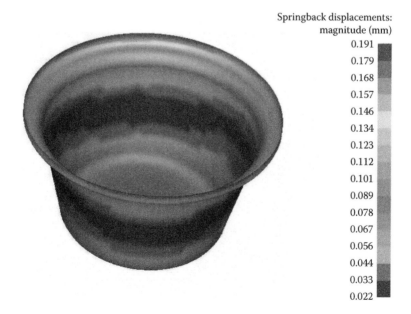

Springback displacements:
magnitude (mm)

0.191
0.179
0.168
0.157
0.146
0.134
0.123
0.112
0.101
0.089
0.078
0.067
0.056
0.044
0.033
0.022

FIGURE 11.3
Punch plate—springback displacement magnitude in experiment 2.

Springback displacements:
magnitude (mm)

| 0.218 |
| 0.205 |
| 0.191 |
| 0.178 |
| 0.165 |
| 0.151 |
| 0.138 |
| 0.125 |
| 0.111 |
| 0.098 |
| 0.085 |
| 0.071 |
| 0.058 |
| 0.044 |
| 0.031 |
| 0.018 |

FIGURE 11.4
Punch plate—springback displacement magnitude in experiment 3.

Springback displacements:
magnitude (mm)

| 0.184 |
| 0.173 |
| 0.163 |
| 0.152 |
| 0.142 |
| 0.131 |
| 0.121 |
| 0.111 |
| 0.100 |
| 0.090 |
| 0.079 |
| 0.069 |
| 0.058 |
| 0.048 |
| 0.037 |
| 0.027 |

FIGURE 11.5
Punch plate—springback displacement magnitude in experiment 4.

FIGURE 11.6
Punch plate—springback displacement magnitude in experiment 5.

FIGURE 11.7
Punch plate—springback displacement magnitude in experiment 6.

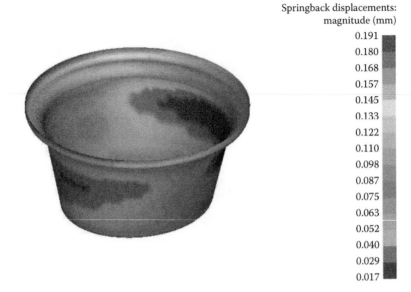

Springback displacements:
magnitude (mm)

0.191
0.180
0.168
0.157
0.145
0.133
0.122
0.110
0.098
0.087
0.075
0.063
0.052
0.040
0.029
0.017

FIGURE 11.8
Punch plate—springback displacement magnitude in experiment 7.

Springback displacements:
magnitude (mm)

0.182
0.172
0.161
0.150
0.139
0.128
0.118
0.107
0.096
0.085
0.074
0.064
0.053
0.042
0.031
0.021

FIGURE 11.9
Punch plate—springback displacement magnitude in experiment 8.

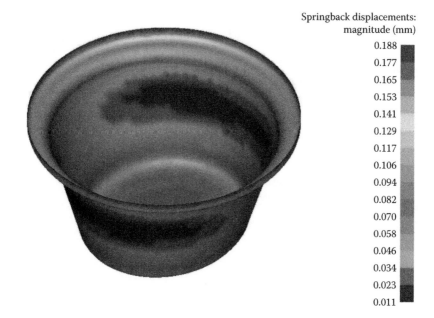

Springback displacements:
magnitude (mm)

0.188	
0.177	
0.165	
0.153	
0.141	
0.129	
0.117	
0.106	
0.094	
0.082	
0.070	
0.058	
0.046	
0.034	
0.023	
0.011	

FIGURE 11.10

Punch plate—springback displacement magnitude in experiment 9.

TABLE 11.4

Punch Plate—Springback Range

Experiment 1	0.01–0.16 mm	Experiment 2	0.02–0.19 mm	Experiment 3	0.01–0.21 mm
Experiment 4	0.02–0.18 mm	Experiment 5	0.01–0.18 mm	Experiment 6	0.009–0.18 mm
Experiment 7	0.01–0.19 mm	Experiment 8	0.02–0.18 mm	Experiment 9	0.01–0.18 mm

11.7 Analysis of Variance

The springback at various cross sections during all nine experiments is recorded and presented in Table 11.5. The average spring back is calculated with actual measurements at various cross sections. For analysis of variance the quality characteristic selected is springback. The S/N ratios are calculated for the quality characteristic, where the smaller the ratio, the better the quality.

The maximum springback displacement magnitude is 0.16 mm and the minimum is 0.02 mm. The maximum springback displacement of 0.085 mm is observed in experiment 8. The results of the analysis of variance are presented in Tables 11.6 and 11.7. The mean S/N ratios are calculated for all

TABLE 11.5

Punch Plate—S/N in Nine Experiments for Springback

Experiment No.	Springback Displacement Magnitude [mm]								Average Displacement [mm]	S/N Ratios
1	0.11	0.12	0.13	0.03	0.04	0.05	0.06	0.07	0.0745	22.55
2	0.10	0.11	0.12	0.03	0.04	0.06	0.07	0.08	0.0767	22.29
3	0.11	0.13	0.14	0.03	0.04	0.06	0.07	0.10	0.0845	21.46
4	0.12	0.13	0.14	0.04	0.05	0.06	0.07	0.08	0.0855	21.36
5	0.02	0.04	0.05	0.06	0.07	0.12	0.13	0.14	0.0786	22.08
6	0.02	0.03	0.04	0.06	0.07	0.08	0.12	0.14	0.0696	23.14
7	0.05	0.06	0.08	0.09	0.10	0.03	0.04	0.15	0.0736	22.65
8	0.12	0.13	0.16	0.03	0.04	0.05	0.06	0.07	0.0838	21.52
9	0.11	0.12	0.13	0.02	0.03	0.05	0.06	0.08	0.0743	22.57

TABLE 11.6

Punch Plate—S/N Ratios at Three Levels for Springback

Parameter	Level	Experiments	Mean S/N Ratio
Blank-holder force [BHF]	1	1, 2, 3	22.10
	2	4, 5, 6	22.19
	3	7, 8, 9	22.25
Coefficient of friction [μ]	1	1, 4, 7	22.35
	2	2, 5, 8	21.97
	3	3, 6, 9	22.39
Die profile radius [R_D]	1	1, 6, 8	22.40
	2	2, 4, 9	22.07
	3	3, 5, 7	22.07
Punch nose radius [R_P]	1	1, 5, 9	22.40
	2	2, 6, 7	22.70
	3	3, 4, 8	21.45

TABLE 11.7

Punch Plate—ANOVA Results for Springback

	BHF	Friction	R_D	R_P
1	22.10	22.35	22.40	22.40
2	22.19	21.97	22.07	22.70
3	22.25	22.39	22.07	21.45
Range	0.14	0.42	0.33	1.25
Rank	4	2	3	1

parameters—blank-holder force, coefficient of friction, die profile radius, and punch nose radius—at all three levels, that is, low, medium, and high. The range is defined as the difference between the maximum and minimum values of S/N ratios for a particular parameter.

Table 11.7 shows the rearrangement of S/N ratios for all variables at all levels. The rank indicates the influence of the input parameter on the output quality characteristic. The result of the orthogonal array indicates that the punch nose radius has a major influence on springback displacement magnitude. Friction is second in rank, the die profile radius is third, and the blank-holder force has the least influence on springback displacement.

11.8 Objective Function Formulation

Linear mathematical relations have been developed from the results of Taguchi design of experiments and analysis of variance between input parameters like blank-holder force, friction coefficient, die profile radius, and punch nose radius. The performance characteristics applied for punch plate is springback. The relationship is presented as follows. Minitab has been used for regression analysis.

The objective function is

Minimize springback

$$\text{Springback} = 0.0488 - (0.000133 * \text{BHF}) - (0.0167 * \mu)$$
$$+ (0.00150 * R_D) + (0.00217 * R_P)$$

Subject to

$$1.2 \le \beta \le 2.2$$
$$3R_D \le R_P \le 6R_D$$
$$F_{d\ max} \le \pi d_m S_0 S_u$$
$$R_D \ge 0.035 \left[50 + (d_0 - d_1) \right] \sqrt{S_0}$$

11.9 Results

The CS algorithm was applied for optimization of springback. The parameters selected for the algorithm are presented in Table 11.8.

During the minimization process, the punch plate diameter, die profile radius, and coefficient of friction were selected as variables. Table 11.9

TABLE 11.8

Cuckoo Search Parameters

Cuckoo Parameters	Set Value
Number of iterations	2000
Number of nests	25
Discovery rate of alien eggs	0.25

TABLE 11.9

Optimization Results—Cuckoo Search

Parameter	Lower Bound	Upper Bound	Optimum
Punch plate diameter [mm]	105	109	109
Corner radius [mm]	4	6	4
Coefficient of friction	0.005	0.15	0.15
Optimized springback displacement magnitude			0.066 mm

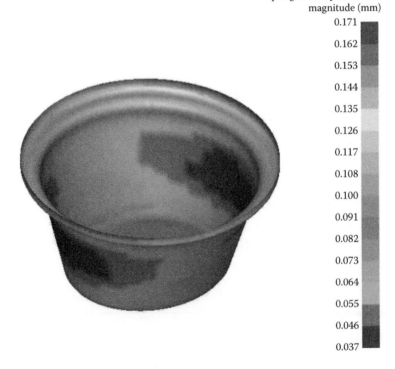

Springback displacements:
magnitude (mm)

0.171
0.162
0.153
0.144
0.135
0.126
0.117
0.108
0.100
0.091
0.082
0.073
0.064
0.055
0.046
0.037

FIGURE 11.11

Punch plate—optimum springback displacement magnitude.

presents results of optimization and optimized value of variables with the lower and upper limits.

The optimum springback obtained with CS is 0.066 mm. The optimum diameter is 109 mm and the die profile radius is 0.4 mm. The coefficient of friction needs to be maintained at an optimum value of 0.15.

The numerical simulation was carried out with the optimum parameters achieved after optimization with CS. The springback measured at different sections of the punch plate is presented in Figures 11.11 through 11.15. The average springback in all sections is measured and observed to be 0.060 mm. This indicates that if optimum parameters are selected, springback can be reduced.

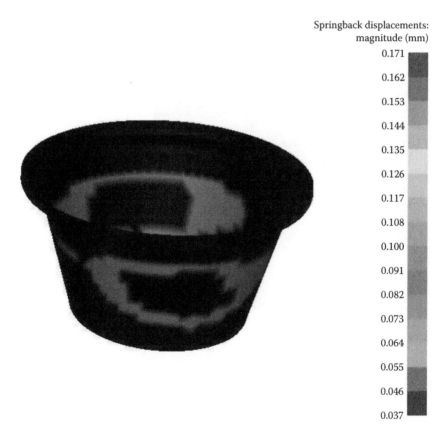

Springback displacements: magnitude (mm)

0.171
0.162
0.153
0.144
0.135
0.126
0.117
0.108
0.100
0.091
0.082
0.073
0.064
0.055
0.046
0.037

FIGURE 11.12
Punch plate—various ranges of optimum springback displacement magnitude.

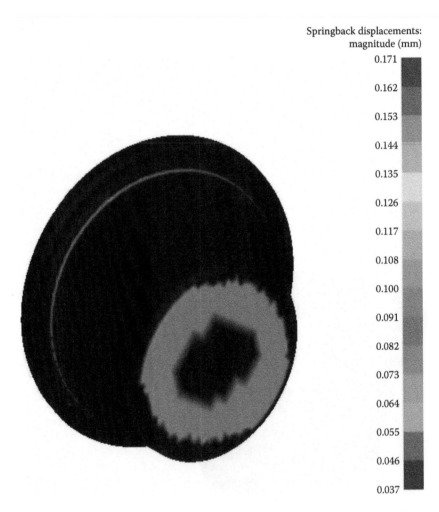

FIGURE 11.13
Punch plate—various ranges of optimum springback displacement magnitude.

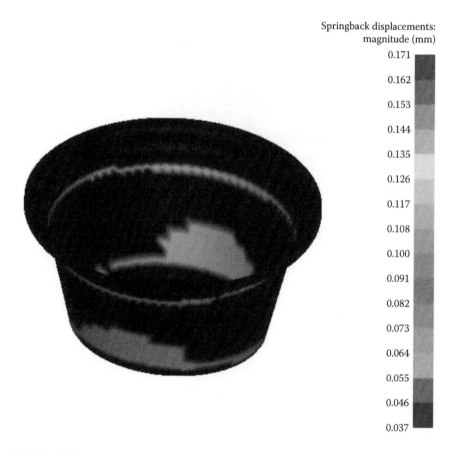

Springback displacements:
magnitude (mm)

0.171
0.162
0.153
0.144
0.135
0.126
0.117
0.108
0.100
0.091
0.082
0.073
0.064
0.055
0.046
0.037

FIGURE 11.14
Punch plate—various ranges of optimum springback displacement magnitude.

Springback displacements:
magnitude (mm)

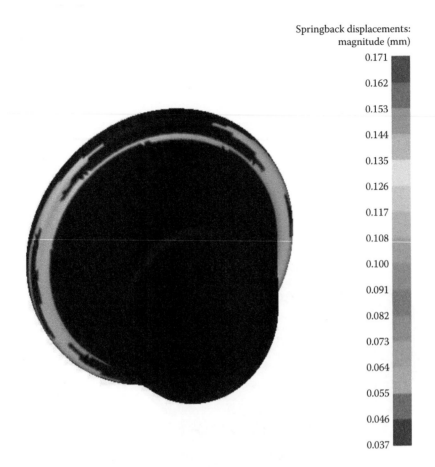

0.171
0.162
0.153
0.144
0.135
0.126
0.117
0.108
0.100
0.091
0.082
0.073
0.064
0.055
0.046
0.037

FIGURE 11.15
Punch plate—various ranges of optimum springback displacement magnitude.

Bibliography

Barthelemy, P., J. Bertolotti, and D. S. Wiersma, A levy flight for light, *Nature* 453 (March 2008), 495–498, DOI: 10.1038/nature06948.

Brown, C. T., L. S. Liebovitch, and R. Glendon, Levy flights in Dobe Ju/'hoansi foraging patterns, *Human Ecology* 35(2) (2007), 129–138. DOI: 10.1007/s10745-006-9083-4.

Fister Jr., I., X.-S. Yang, D. Fister, and I. Fister, Cuckoo search: A brief literature review, in: *Cuckoo Search and Firefly Algorithm: Theory and Applications* (Editor X.-S. Yang), Springer Publications, New York, pp. 49–62, 2014.

Gandomi, A. H., X. S. Yang, and A. H. Alavi, Cuckoo search algorithm: A meta-heuristic approach to solve structural optimization problems, *Engineering with Computers* 29(1) (January 2013), 17–35, DOI: 10.1007/s00366-011-0241-y.

Layeb, A., Novel quantum-inspired cuckoo search for Knapsack problems, *International Journal of Bioinspired Computation* 3(5) (September 2011), 297–305, DOI: 10.1504/IJBIC.2011.042260.

Pavlyukevich, I., Levy flights, non-local search and simulated annealing, *Journal of Computational Physics* 226(2) (October 2007), 1830–1844, DOI: 10.1016/j.jcp.2007.06.008.

Payne, R. B., M. D. Sorenson, and K. Klitz, *The Cuckoos*, Oxford University Press, Oxford, U.K., 2005.

Appendix A

A.1 Sample Code for Cohort Intelligence

```
clc
clf('reset')
clear all
clear fig
format long
n= 3;%input('ENTER THE NO OF VARIABLE: ');
a= [57 1.5 0.005];%input('Enter the lower bound= ');
b= [100 100 0.15];%input('Enter the upper bound= ');
c= 6;%input('Enter number of candidates= ');
r= 0.96;%input('Enter Value of R from 0 to 1=');
e= 1e-6;%input('Enter convergence value');
N=[1 2];
iter=100; rnge=ones(c,1)*(b-a); index=1; t=2000; rnge=rnge/2;
lo=ones(c,1)*a;
up=ones(c,1)*b;
a=ones(c,1)*a;
b=ones(c,1)*b;

tot=0;
minium=5e5;
cur_min=5e5;
F=[1];
g=((b-a)*r)/2;
tic;
iter_no = 1;
entcov=0;
extcov=0;
figure(1);
        title('Convergence of Solutions');
        xlabel('Iterations');
        ylabel('F(x)');
while(iter_no<t)
cur_min=8e8;
    for j=1:c               %calculate value of each candidate
        for m=1:iter            %loop to generate t solutions

            for k=1:n       %random value generation for every
                            variable
```

```
            x(k) = lo(j,k) + (up(j,k) - lo(j,k)).*rand(1,1);
        end
        for k=1:n
            V(m,k) = x(k);
        end
        G=sealingcovercon(x);

        [~, d2]=size(G);
        for i=1:d2
            if (G(i)<=0)
                u(i)=0;
            else
                u(i)=1e20;
            end
        end
        sh(m)=sealingcover(G,u,x);
    end

    [F(j),m] = min(sh);              %finding best solution
    solall(iter_no,j) = F(j);        %This is not as per
                                        the algorithm.

    if(cur_min>F(j))

        cur_min=F(j);

        if(minium>cur_min)
            minium = cur_min;
        end
    end

    F(j)=1/F(j);
    SV(j,:)=V(m,:);
    tot=tot+F(j);
end

solmi(iter_no,:) = min(solall(iter_no,:));
solmx(iter_no,:) = max(solall(iter_no,:));

for j=1:c                                %finding probability
    prob(j)=F(j)/tot;
end
prob=abs(prob);
for j=1:c                       %selecting candidate to follow
    N(j)=RouletteWheelSelection(prob);
end

for k=1:c                   %to find out the lower bound and
                                    upper
    for j=1:n                %bound and update the bound as per
```

```
            lo(k,j) = SV(N(k),j) - rnge(k,j);        %candidates
                                                        selected in
                                                        the roulette
            if lo(k,j) < a(k,j)              %wheel.
                lo(k,j) = a(k,j);
            end
            up(k,j) = SV(N(k),j) + rnge(k,j);
            if up(k,j) > b(k,j)
                up(k,j) - b(k,j),
            end
        end
    end
    if (iter_no>=2)         % This is the convergence criteria
        difmx(iter_no-1) = solmx(iter_no) - solmx(iter_no-1);
                            % ||Max((FC)n)-Max((FC)n-1)||<=e
        difmx = abs(difmx); %
        difmi(iter_no-1) = solmi(iter_no) -
          solmi(iter_no-1);   % ||Min((FC)n)-Min((FC)
                                    n-1)||<=e
        difmi = abs(difmi); %
        difmnx(iter_no) = solmx(iter_no) -
          solmi(iter_no);       % ||Max((FC)n)-Min((FC)n)||<=e
        difmnx = abs(difmnx);

        if (difmx(iter_no-1)<e) && (difmi(iter_no-1)<e) &&
          (difmnx(iter_no-1)<e && extcov==3)
            break;
        end

      if (difmx(iter_no-1)<e) && (difmi(iter_no-1)<e) &&
        (difmnx(iter_no-1)<e && entcov==0)
            entcov=entcov+1;
          %best(runs)=min_now;
            prev(index,:)=SV(index,:);
          rnge=ones(c,n);
            for k=1:c
                    for j=1:n
                        lo(k,j)=prev(index,j) - rnge(k,j);
                        if lo(k,j)<lo
                            lo(k,j)=lo;
                        end
                        up(k,j)=prev(index,j) + rnge(k,j);
                        if up(k,j)>up
                            up(k,j)=up;
                        end
                    end
                end
                extcov=3;
            %break;
        end
    end
end
```

```
    tot=0;
        rnge = rnge*r;
    iter_no = iter_no+1;
end
time=toc;
plot(solall, '--gs','linewidth',2,'MarkerEdgeColor','b','Marker
  FaceColor','g','MarkerSize',10);   % Ploting of solutions
fprintf('\n\nNo. of iterations required are - %d',iter_no);
fprintf('\n\nTime Required = %6f',time);
fprintf('\n\n d1 = %6f',x(1));
fprintf('\n\n r = %6f',x(2));
fprintf('\n\n muu = %6f',x(3));
% fprintf('\n\n Rp = %6f',x(4));
fprintf('\n\nminima is = %e',cur_min);


function y = sealingcover(G,u,x)
y = 0.782+ 0.000778*(((pi/4) *
   ((68^2)-((x(1)+(2*x(2)))^2)))*(2.5/1000))-0.0433*(x(3))- ...
        0.00039*(0.035*(50+(68-x(1))))*(0.8^0.5)+...
        0.00069*((3+(3*rand(1,1)))*(0.035*(50+(68-x(1))))*(0.8^0.5))
           + (u(1) * G(1)^2) + (u(2) * G(2)^2);


function y = Punchplatecon(x)
y(1) = 1.6 - ((0.035*(50+(68-x(1))))*(0.8^0.5));
y(2) = ((0.035*(50+(68-x(1))))*(0.8^0.5)) - 4.8;
```

A.2 Sample Code for Cuckoo Search

```
function [bestsol,fval]=cuckoo_search_spring(time)
format long;
help cuckoo_search_spring.m
if nargin<1,
    % Number of iteraions
    time=2000;
end
disp('Computing ... it may take a few minutes.');
% Number of nests (or different solutions)
n=25;
% Discovery rate of alien eggs/solutions
pa=0.25;
% Simple bounds of the search domain
% Lower bounds and upper bounds
Lb=[105 4 0.005];
Ub=[109 6 0.15];
```

```
% Random initial solutions
for i=1:n,
nest(i,:)=Lb+(Ub-Lb).*rand(size(Lb));
end
% Get the current best
fitness=10^10*ones(n,1);
[fmin,bestnest,nest,fitness]=get_best_nest(nest,nest,fitness);
N_iter=0;
%% Starting iterations
for t=1:time,
    % Generate new solutions (but keep the current best)
     new_nest=get_cuckoos(nest,bestnest,Lb,Ub);
    [fnew,best,nest,fitness]=get_best_nest(nest,new_nest,fitness);
    % Update the counter
      N_iter=N_iter+n;
    % Discovery and randomization
      new_nest=empty_nests(nest,Lb,Ub,pa) ;
    % Evaluate this solution
    [fnew,best,nest,fitness]=get_best_nest(nest,new_nest,fitness);
    % Update the counter again
      N_iter=N_iter+n;
    % Find the best objective so far
    if fnew<fmin,
        fmin=fnew;
        bestnest=best ;
    end
end %% End of iterations
%% Post-optimization processing
%% Display all the nests
disp(strcat('Total number of iterations=',num2str(N_iter)));
fmin
bestnest
%%---------------All subfunctions are list below------------------
%% Get cuckoos by ramdom walk


function nest=get_cuckoos(nest,best,Lb,Ub)
% Levy flights
n=size(nest,1);
beta=3/2;
sigma=(gamma(1+beta)*sin(pi*beta/2)/(gamma((1+beta)/2)*beta*2^
  ((beta-1)/2)))^(1/beta);
for j=1:n,
    s=nest(j,:);
    % This is a simple way of implementing Levy flights
    % For standard random walks, use step=1;
    %% Levy flights by Mantegna's algorithm
    u=randn(size(s))*sigma;
    v=randn(size(s));
    step=u./abs(v).^(1/beta);
```

```
       % In the next equation, the difference factor (s-best)
         means that
       % when the solution is the best solution, it remains
         unchanged.
       stepsize=0.01*step.*(s-best);
       % Here the factor 0.01 comes from the fact that L/100
         should the typical
       % step size of walks/flights where L is the typical
         lenghtscale;
       % otherwise, Levy flights may become too aggresive/
         efficient,
       % which makes new solutions (even) jump out side of the
         design domain
       % (and thus wasting evaluations).
       % Now the actual random walks or flights
       s=s+stepsize.*randn(size(s));
     % Apply simple bounds/limits
     nest(j,:)=simplebounds(s,Lb,Ub);
end
%% Find the current best nest


function
  [fmin,best,nest,fitness]=get_best_nest(nest,newnest,fitness)
% Evaluating all new solutions
for j=1:size(nest,1),
    fnew=fobj(newnest(j,:));
    if fnew<=fitness(j),
        fitness(j)=fnew;
        nest(j,:)=newnest(j,:);
    end
end
% Find the current best
[fmin,K]=min(fitness) ;
best=nest(K,:);
%% Replace some nests by constructing new solutions/nests


function new_nest=empty_nests(nest,Lb,Ub,pa)
% A fraction of worse nests are discovered with a probability pa
n=size(nest,1);
% Discovered or not -- a status vector
K=rand(size(nest))>pa;
% In real world, if a cuckoo's egg is very similar to a host's
  eggs, then
% this cuckoo's egg is less likely to be discovered, thus the
  fitness should
% be related to the difference in solutions.  Therefore, it is
  a good idea
% to do a random walk in a biased way with some random step sizes.
```

```
nestn1=nest(randperm(n),:);
nestn2=nest(randperm(n),:);
%% New solution by biased/selective random walks
stepsize=rand*(nestn1-nestn2);
new_nest=nest+stepsize.*K;
for j=1:size(new_nest,1)
    s=new_nest(j,:);
  new_nest(j,:)=simplebounds(s,Lb,Ub);
end
% Application of simple constraints
function s=simplebounds(s,Lb,Ub)
  % Apply the lower bound
  ns_tmp=s;
  I=ns_tmp<Lb;
  ns_tmp(I)=Lb(I);
  % Apply the upper bounds
  J=ns_tmp>Ub;
  ns_tmp(J)=Ub(J);
  % Update this new move
  s=ns_tmp;

function z=fobj(x)
% The well-known spring design problem
z=0.0488 - (0.000133*(((pi/4) * ((150^2)-((x(1)+
  (2*x(2)))^2)))*(2.5/1000))) - (0.0167* x(3)) +...
    (0.00150*(0.035*(50+(150-x(1))))*(0.8^0.5))...
    + (0.00217*((3+(3*rand(1,1)))*
      (0.035*(50+(150-x(1))))*(0.8^0.5)));
z=z+getnonlinear(x);

function Z=getnonlinear(x)
Z=0;
% Penalty constant
lam=10^15;
% Inequality constraints
g(1) = 2.5 - ((0.035*(50+(150-x(1))))*(0.8^0.5));
g(2) = ((0.035*(50+(150-x(1))))*(0.8^0.5)) -8;
% No equality constraint in this problem, so empty;
geq=[];
% Apply inequality constraints
for k=1:length(g),
    Z=Z+ lam*g(k)^2*getH(g(k));
end
% Apply equality constraints
for k=1:length(geq),
  Z=Z+lam*geq(k)^2*getHeq(geq(k));
end
% Test if inequalities hold
% Index function H(g) for inequalities
```

```
function H=getH(g)
if g<=0,
    H=0;
else
    H=1;
end
% Index function for equalities

function H=getHeq(geq)
if geq==0,
    H=0;
else
    H=1;
end
% ---------------- end ------------------------------
```

A.3 Sample Code for Fire Fly Algorithm

```
function fa_mincon
% parameters [n N_iteration alpha betamin gamma]
para=[40 500 0.5 0.2 1];
format long;
% Simple bounds/limits
disp('Solve the simple spring design problem ...');
Lb=[46 2.5 0.005];
Ub=[50 3.5 0.15];
% Initial random guess
u0=Lb+(Ub-Lb).*rand(size(Lb));
[u,fval,NumEval]=ffa_mincon(@cost,@constraint,u0,Lb,Ub,para);
% Display results
bestsolution=u
bestojb=fval
total_number_of_function_evaluations=NumEval
%% Objective function
 function z=cost(x)
z = 349 - (157*(((pi/4) * ((65^2)-((x(1)+(2*x(2)))^2)))*
    (2.5/1000))) + (283 * x(3)) + ...
    (337*(0.035*(50+(65-x(1))))*(1^0.5))...
    + (61*((3+(3*rand(1,1)))*(0.035*(50+(65-x(1))))*(1^0.5)));

function [g,geq]=constraint(x)
% All nonlinear inequality constraints should be here
% If no inequality constraint at all, simple use g=[];
g(1) = 2.5 - ((0.035*(50+(150-x(1))))*(0.8^0.5));
g(2) = ((0.035*(50+(150-x(1))))*(0.8^0.5)) - 8;
```

```
% If no equality constraint at all, put geq=[] as follows
geq=[];

% Start FA
function [nbest,fbest,NumEval]...
          =ffa_mincon(fhandle,nonhandle,u0, Lb, Ub, para)
% Check input parameters (otherwise set as default values)
if nargin<6, para=[20 50 0.25 0.20 1]; end
if narqin<5, Ub=[]; end
if nargin<4, Lb=[]; end
if nargin<3,
disp('Usuage: FA_mincon(@cost, @constraint,u0,Lb,Ub,para)');
end
% n=number of fireflies
% MaxGeneration=number of pseudo time steps
% ------------------------------------------------
% alpha=0.25;      % Randomness 0--1 (highly random)
% betamn=0.20;     % minimum value of beta
% gamma=1;         % Absorption coefficient
% ------------------------------------------------
n=para(1);   MaxGeneration=para(2);
alpha=para(3); betamin=para(4); gamma=para(5);
% Total number of function evaluations
NumEval=n*MaxGeneration;
% Check if the upper bound & lower bound are the same size
if length(Lb) ~=length(Ub),
    disp('Simple bounds/limits are improper!');
    return
end
% Calcualte dimension
d=length(u0);
% Initial values of an array
zn=ones(n,1)*10^100;
% ------------------------------------------------
% generating the initial locations of n fireflies
[ns,Lightn]=init_ffa(n,d,Lb,Ub,u0);
% Iterations or pseudo time marching
for k=1:MaxGeneration,     %%%% start iterations
% This line of reducing alpha is optional
 alpha=alpha_new(alpha,MaxGeneration);
% Evaluate new solutions (for all n fireflies)
for i=1:n,
   zn(i)=Fun(fhandle,nonhandle,ns(i,:));
   Lightn(i)=zn(i);
end

% Ranking fireflies by their light intensity/objectives
[Lightn,Index]=sort(zn);
ns_tmp=ns;
```

```
for i=1:n,
 ns(i,:)=ns_tmp(Index(i),:);
end

%% Find the current best
nso=ns; Lighto=Lightn;
nbest=ns(1,:); Lightbest=Lightn(1);

% For output only
fbest=Lightbest;

% Move all fireflies to the better locations
[ns]=ffa_move(n,d,ns,Lightn,nso,Lighto,nbest,...
      Lightbest,alpha,betamin,gamma,Lb,Ub);

end

% ---------------------------------------------------------
% ----- All the subfunctions are listed here ------------
% The initial locations of n fireflies
function [ns,Lightn]=init_ffa(n,d,Lb,Ub,u0)
  % if there are bounds/limits,
if length(Lb)>0,
   for i=1:n,
   ns(i,:)=Lb+(Ub-Lb).*rand(1,d);
   end
else
   % generate solutions around the random guess
   for i=1:n,
   ns(i,:)=u0+randn(1,d);
   end
end

% initial value before function evaluations
Lightn=ones(n,1)*10^100;

% Move all fireflies toward brighter ones
function [ns]=ffa_move(n,d,ns,Lightn,nso,Lighto,...
              nbest,Lightbest,alpha,betamin,gamma,Lb,Ub)
% Scaling of the system
scale=abs(Ub-Lb);

% Updating fireflies
for i=1:n,
% The attractiveness parameter beta=exp(-gamma*r)
   for j=1:n,
      r=sqrt(sum((ns(i,:)-ns(j,:)).^2));
      % Update moves
```

```
if Lightn(i)>Lighto(j), % Brighter and more attractive
    beta0=1; beta=(beta0-betamin)*exp(-gamma*r.^2)+betamin;
    tmpf=alpha.*(rand(1,d)-0.5).*scale;
    ns(i,:)=ns(i,:).*(1-beta)+nso(j,:).*beta+tmpf;
        end
    end % end for j

end % end for i

% Check if the updated solutions/locations are within limits
[ns]=findlimits(n,ns,Lb,Ub);

function alpha=alpha_new(alpha,NGen)
% alpha_n=alpha_0(1-delta)^NGen=10^(-4);
% alpha_0=0.9
delta=1-(10^(-4)/0.9)^(1/NGen);
alpha=(1-delta)*alpha;

% Make sure the fireflies are within the bounds/limits
function [ns]=findlimits(n,ns,Lb,Ub)
for i=1:n,
      % Apply the lower bound
  ns_tmp=ns(i,:);
  I=ns_tmp<Lb;
  ns_tmp(I)=Lb(I);

  % Apply the upper bounds
  J=ns_tmp>Ub;
  ns_tmp(J)=Ub(J);
  % Update this new move
  ns(i,:)=ns_tmp;
end

% -----------------------------------------
% d-dimensional objective function
function z=Fun(fhandle,nonhandle,u)
% Objective
z=fhandle(u);

% Apply nonlinear constraints by the penalty method
% Z=f+sum_k=1^N lam_k g_k^2 *H(g_k) where lam_k >> 1
z=z+getnonlinear(nonhandle,u);

function Z=getnonlinear(nonhandle,u)
Z=0;
% Penalty constant >> 1
lam=10^15; lameq=10^15;
% Get nonlinear constraints
[g,geq]=nonhandle(u);
```

```
% Apply inequality constraints as a penalty function
for k=1:length(g),
    Z=Z+ lam*g(k)^2*getH(g(k));
end
% Apply equality constraints (when geq=[], length->0)
for k=1:length(geq),
    Z-Z+lameq*geq(k)^2*geteqH(geq(k));
end

% Test if inequalities hold
% H(g) which is something like an index function
function H=getH(g)
if g<=0,
    H=0;
else
    H=1;
end

% Test if equalities hold
function H=geteqH(g)
if g==0,
    H=0;
else
    H=1;
end
```

A.4 Sample Code for Flower Pollination Algorithm

```
function [best,fmin,N_iter]=fpa(para)
% Default parameters
if nargin<1,
    para=[20 0.8];
end
n=para(1);              % Population size, typically 10 to 25
p=para(2);              % probability switch

% Iteration parameters
N_iter=2000;            % Total number of iterations

% Dimension of the search variables
d=3;
Lb=[57 1.5 0.005];
Ub=[61 2.5 0.15];

% Initialize the population/solutions
for i=1:n,
```

```
  Sol(i,:)=Lb+(Ub-Lb).*rand(1,d);
  Fitness(i)=Fun(Sol(i,:));
end

% Find the current best
[fmin,I]=min(Fitness);
best=Sol(I,:);
S=Sol;

% Start the iterations -- Flower Algorithm
for t=1:N_iter,
        % Loop over all bats/solutions
        for i=1:n,
          % Pollens are carried by insects and thus can move in
          % large scale, large distance.
          % This L should replace by Levy flights
          % Formula: x_i^{t+1}=x_i^t+ L (x_i^t-gbest)
          if rand>p,
          %% L=rand;
          L=Levy(d);
          dS=L.*(Sol(i,:)-best);
          S(i,:)=Sol(i,:)+dS;

          % Check if the simple limits/bounds are OK
          S(i,:)=simplebounds(S(i,:),Lb,Ub);

          % If not, then local pollenation of neighbor flowers
          else
              epsilon=rand;
              % Find random flowers in the neighbourhood
              JK=randperm(n);
              % As they are random, the first two entries also
                 random
              % If the flower are the same or similar species,
                 then
              % they can be pollenated, otherwise, no action.
              % Formula: x_i^{t+1}+epsilon*(x_j^t-x_k^t)
              S(i,:)=S(i,:)+epsilon*(Sol(JK(1),:)-Sol(JK(2),:));
              % Check if the simple limits/bounds are OK
              S(i,:)=simplebounds(S(i,:),Lb,Ub);
          end

          % Evaluate new solutions
           Fnew=Fun(S(i,:));
          % If fitness improves (better solutions found),
            update then
            if (Fnew<=Fitness(i)),
                Sol(i,:)=S(i,:);
                Fitness(i)=Fnew;
            end
```

```
            % Update the current global best
            if Fnew<=fmin,
                    best=S(i,:)    ;
                    fmin=Fnew    ;
            end
        end
        % Display results every 100 iterations
        if round(t/100)==t/100,
        best
        fmin
        end

end


% Output/display
disp(['Total number of evaluations: ',num2str(N_iter*n)]);
disp(['Best solution=',num2str(best),' fmin=',num2str(fmin)]);
% Application of simple constraints

function s=simplebounds(s,Lb,Ub)
  % Apply the lower bound
  ns_tmp=s;
  I=ns_tmp<Lb;
  ns_tmp(I)=Lb(I);
  % Apply the upper bounds
  J=ns_tmp>Ub;
  ns_tmp(J)=Ub(J);
  % Update this new move
  s=ns_tmp;
% Draw n Levy flight sample

function L=Levy(d)
% Levy exponent and coefficient
% For details, see Chapter 11 of the following book:
% Xin-She Yang, Nature-Inspired Optimization Algorithms,
  Elsevier, (2014).
beta=3/2;
sigma=(gamma(1+beta)*sin(pi*beta/2)/(gamma((1+beta)/2)*beta*2^
  ((beta-1)/2)))^(1/beta);
    u=randn(1,d)*sigma;
    v=randn(1,d);
    step=u./abs(v).^(1/beta);
L=0.01*step;


function z=Fun(x)
z= 0.0239 + (0.00159*(((pi/4) * ((68^2)-((x(1)+(
  2*x(2)))^2)))*(2.5/1000))) - (0.0324* x(3)) -...
    (0.00144*(0.035*(50+(68-x(1))))*(0.8^0.5))...
```

```
     + (0.00067*((3+(3*rand(1,1)))*(0.035*(50+(68-x(1))))*
        (0.8^0.5)));
z=z+getnonlinear(x);

function Z=getnonlinear(x)
Z=0;
% Penalty constant
lam=1000;

% Inequality constraints
g(1) = 1.6 - ((0.035*(50+(68-x(1))))*(0.8^0.5));
g(2) = ((0.035*(50+(68-x(1))))*(0.8^0.5)) - 4.8;

% No equality constraint in this problem, so empty;
geq=[];

% Apply inequality constraints
for k=1:length(g),
    Z=Z+ lam*g(k)^2*getH(g(k));
end
% Apply equality constraints
for k=1:length(geq),
   Z=Z+lam*geq(k)^2*getHeq(geq(k));
end
% Test if inequalities hold
% Index function H(g) for inequalities
function H=getH(g)
if g<=0,
    H=0;
else
    H=1;
end
% Index function for equalities
function H=getHeq(geq)
if geq==0,
   H=0;
else
   H=1;
end
% ---------------- end ------------------------
```

A.5 Sample Code for Grey Wolf Optimization

```
clear all
close all
clc
clf
```

```
format long
SearchAgents_no=25; % Number of search agents
Function_name='F25'; % Name of the function
Max_iteration=2000; % Maximum number of iterations
% Load details of the selected benchmark function
[lb,ub,dim,fobj]=Get_Functions_details(Function_name);
[Best_score,Best_pos,GWO_cg_curve]=GWO(SearchAgents_no,Max_
   iteration,lb,ub,dim,fobj);
display(['The best solution obtained by GWO is : ',
   num2str(Best_pos)]);
display(['The best optimal value of the objective funciton
   found by GWO is : ', num2str(Best_score)]);

function [lb,ub,dim,fobj] = Get_Functions_details(F)
switch F
    case 'F25'
        fobj = @F25;
        lb=[105 4 0.005];
        ub=[109 6 0.15];
        dim=3;
end
end
function o = F25(x)
   z = 0.795 - (0.000133*(((pi/4) * ((150^2)-((x(1)+
(2*x(2)))^2)))*(2.5/1000))) + (0.243 * x(3)) - ...
    (0.00033*(0.035*(50+(150-x(1))))*(0.8^0.5))...
    - (0.00217*((3+(3*rand(1,1)))*(0.035*(50+(150-x(1))))*(0.8^0.5)));
o = z+getnonlinear(x);
end

function Z=getnonlinear(x)
Z=0;
% Penalty constant
lam=100;
% Inequality constraints
g(1) = 2.5 - ((0.035*(50+(150-x(1))))*(0.8^0.5));
g(2) = ((0.035*(50+(150-x(1))))*(0.8^0.5)) - 8;
% No equality constraint in this problem, so empty;
geq=[];
% Apply inequality constraints
for k=1:length(g),
    Z=Z + lam*g(k)^2*getH(g(k));
end
% Apply equality constraints
for k=1:length(geq),
   Z=Z+lam*geq(k)^2*getHeq(geq(k));
end
end
% Test if inequalities hold
% Index function H(g) for inequalities
```

```
function H=getH(g)
if g<=0,
    H=0;
else
    H=1;
end
end
% Index function for equalities
function H=getHeq(qeq)
if geq==0,
    H=0;
else
    H=1;
end
end


function [Alpha_score,Alpha_pos,Convergence_curve]=
GWO(SearchAgents_no,Max_iter,lb,ub,dim,fobj)
% initialize alpha, beta, and delta_pos
Alpha_pos=zeros(1,dim);
Alpha_score=inf; %change this to -inf for maximization problems
Beta_pos=zeros(1,dim);
Beta_score=inf; %change this to -inf for maximization problems
Delta_pos=zeros(1,dim);
Delta_score=inf; %change this to -inf for maximization problems
%Initialize the positions of search agents
Positions=initialization(SearchAgents_no,dim,ub,lb);
Convergence_curve=zeros(1,Max_iter);
l=0;% Loop counter
% Main loop
while l<Max_iter
    for i=1:size(Positions,1)
        % Return back the search agents that go beyond the
        boundaries of the search space
        Flag4ub=Positions(i,:)>ub;
        Flag4lb=Positions(i,:)<lb;
        Positions(i,:)=(Positions(i,:).*(~(Flag4ub+Flag4lb)))+
            ub.*Flag4ub+lb.*Flag4lb;
        % Calculate objective function for each search agent
        fitness=fobj(Positions(i,:));
        % Update Alpha, Beta, and Delta
        if fitness<Alpha_score
            Alpha_score=fitness; % Update alpha
            Alpha_pos=Positions(i,:);
        end
        if fitness>Alpha_score && fitness<Beta_score
            Beta_score=fitness; % Update beta
            Beta_pos=Positions(i,:);
        end
```

```
        if fitness>Alpha_score && fitness>Beta_score &&
          fitness<Delta_score
              Delta_score=fitness; % Update delta
              Delta_pos=Positions(i,:);
          end
     end
     a=2-1*((2)/Max_iter); % a decreases linearly fron 2 to 0
     % Update the Position of search agents including omegas
     for i=1:size(Positions,1)
         for j=1:size(Positions,2)
             r1=rand(); % r1 is a random number in [0,1]
             r2=rand(); % r2 is a random number in [0,1]
             A1=2*a*r1-a;
             C1=2*r2;
             D_alpha=abs(C1*Alpha_pos(j)-Positions(i,j));
             X1=Alpha_pos(j)-A1*D_alpha;
             r1=rand();
             r2=rand();
             A2=2*a*r1-a;
             C2=2*r2;
             D_beta=abs(C2*Beta_pos(j)-Positions(i,j));
             X2=Beta_pos(j)-A2*D_beta;
             r1=rand();
             r2=rand();
             A3=2*a*r1-a;
             C3=2*r2;

             D_delta=abs(C3*Delta_pos(j)-Positions(i,j));
             X3=Delta_pos(j)-A3*D_delta;
             Positions(i,j)=(X1+X2+X3)/3;
         end
     end
     l=l+1;
     Convergence_curve(l)=Alpha_score;
End


function Positions=initialization(SearchAgents_no,dim,ub,lb)
Boundary_no= size(ub,2);
% If the boundaries of all variables are equal and user enter
  a signle
% number for both ub and lb
if Boundary_no==1
    Positions=rand(SearchAgents_no,dim).*(ub-lb)+lb;
end
% If each variable has a different lb and ub
if Boundary_no>1
    for i=1:dim
```

```
        ub_i=ub(i);
        lb_i=lb(i);
        Positions(:,i)=rand(SearchAgents_no,1).*(ub_i-lb_i)+lb_i;
    end
end
```

A.6 Sample Code for Ant Lion Optimization

```
clear all
clc
SearchAgents_no=40; % Number of search agents
Function_name='F1'; % Name of the function
Max_iteration=500; % Maximum number of iterations
% Load details of the selected benchmark function
[lb,ub,dim,fobj]=Get_Functions_details(Function_name);
[Best_score,Best_pos,cg_curve]=ALO(SearchAgents_no,Max_
    iteration,lb,ub,dim,fobj);
display(['The best solution obtained by ALO is : ',
    num2str(Best_pos)]);
display(['The best optimal value of the objective funciton
    found by ALO is : ', num2str(Best_score)]);

function [lb,ub,dim,fobj] = Get_Functions_details(F)
switch F
    case 'F1'
        fobj = @F1;
        lb=[108 2 0.005];
        ub=[112 4 0.15];
        dim=3;
end
end
% F1
function o = F1(x)
z =  0.943 + (0.000667*(((pi/4) * ((160^2)-((x(1)+(
    2*x(2)))^2)))*(2.5/1000))) + (0.0333 * x(3)) + ...
    (0.00333*(0.035*(50+(160-x(1))))*(1.2^0.5))...
    + (0.00167*((3+(3*rand(1,1)))*(0.035*(50+(160-x(1))))*(1.2^0.5)));
o = z+getnonlinear(x);
end
function Z=getnonlinear(x)
Z=0;
% Penalty constant
lam=1000;
```

```
% Inequality constraints
g(1) = 2 - ((0.035*(50+(160-x(1))))*(1.2^0.5));
g(2) = ((0.035*(50+(160-x(1))))*(1.2^0.5)) - 4;
% No equality constraint in this problem, so empty;
geq=[];
% Apply inequality constraints
for k=1:length(g),
    Z=Z + lam*g(k)^2*getH(g(k));
end
% Apply equality constraints
for k=1:length(geq),
    Z=Z+lam*geq(k)^2*getHeq(geq(k));
end
end
% Test if inequalities hold
% Index function H(g) for inequalities
function H=getH(g)
if g<=0,
    H=0;
else
    H=1;
end
end
% Index function for equalities
function H=getHeq(geq)
if geq==0,
    H=0;
else
    H=1;
end
end


function [Elite_antlion_fitness,Elite_antlion_position,
Convergence_curve]=ALO(N,Max_iter,lb,ub,dim,fobj)
% Initialize the positions of antlions and ants
antlion_position=initialization(N,dim,ub,lb);
ant_position=initialization(N,dim,ub,lb);
% Initialize variables to save the position of elite, sorted
  antlions,
% convergence curve, antlions fitness, and ants fitness
Sorted_antlions=zeros(N,dim);
Elite_antlion_position=zeros(1,dim);
Elite_antlion_fitness=inf;
Convergence_curve=zeros(1,Max_iter);
antlions_fitness=zeros(1,N);
ants_fitness=zeros(1,N);
% Calculate the fitness of initial antlions and sort them
for i=1:size(antlion_position,1)
    antlions_fitness(1,i)=fobj(antlion_position(i,:));
end
```

```
[sorted_antlion_fitness,sorted_indexes]=sort(antlions_fitness);
for newindex=1:N
    Sorted_antlions(newindex,:)=
      antlion_position(sorted_indexes(newindex),:);
end
Elite_antlion_position=Sorted_antlions(1,:);
Elite_antlion_fitness=sorted_antlion_fitness(1);
% Main loop start from the second iteration since the
  first iteration
% was dedicated to calculating the fitness of antlions
Current_iter=2;
while Current_iter<Max_iter+1

    % This for loop simulate random walks
    for i=1:size(ant_position,1)
        % Select ant lions based on their fitness (the better
          anlion the higher chance of catching ant)
        Rolette_index=RouletteWheelSelection(1./
          sorted_antlion_fitness);
        if Rolette_index==-1
           Rolette_index=1;
        end
            % RA is the random walk around the selected antlion
              by rolette wheel
        RA=Random_walk_around_antlion(dim,Max_iter,lb,ub,
          Sorted_antlions(Rolette_index,:),Current_iter);
            % RA is the random walk around the elite (best
              antlion so far)
        [RE]=Random_walk_around_antlion(dim,Max_iter,lb,ub,
          Elite_antlion_position(1,:),Current_iter);
          ant_position(i,:)= (RA(Current_iter,:)+RE(Current_
            iter,:))/2; % Equation (2.13) in the paper
    end
  for i=1:size(ant_position,1)
      % Boundar checking (bring back the antlions of ants
        inside search
      % space if they go beyoud the boundaries
      Flag4ub=ant_position(i,:)>ub;
      Flag4lb=ant_position(i,:)<lb;
      ant_position(i,:)=(ant_position(i,:).*(~(Flag4ub+Flag4
        lb)))+ub.*Flag4ub+lb.*Flag4lb;
      ants_fitness(1,i)=fobj(ant_position(i,:));
  end
  % Update antlion positions and fitnesses based of the ants
    (if an ant
  % becomes fitter than an antlion we assume it was cought
    by the antlion
  % and the antlion update goes to its position to build the
    trap)
  double_population=[Sorted_antlions;ant_position];
```

```
double_fitness=[sorted_antlion_fitness ants_fitness];
[double_fitness_sorted I]=sort(double_fitness);
double_sorted_population=double_population(I,:);

antlions_fitness=double_fitness_sorted(1:N);
Sorted_antlions=double_sorted_population(1:N,:);
% Update the position of elite if any antlinons becomes
    fitter than it
if antlions_fitness(1)<Elite_antlion_fitness
    Elite_antlion_position=Sorted_antlions(1,:);
    Elite_antlion_fitness=antlions_fitness(1);
end
% Keep the elite in the population
Sorted_antlions(1,:)=Elite_antlion_position;
antlions_fitness(1)=Elite_antlion_fitness;
% Update the convergence curve
Convergence_curve(Current_iter)=Elite_antlion_fitness;
% Display the iteration and best optimum obtained so far
if mod(Current_iter,50)==0
    display(['At iteration ', num2str(Current_iter), ' the
        elite fitness is ', num2str(Elite_antlion_fitness)]);
end
Current_iter=Current_iter+1;
End


function X=initialization(SearchAgents_no,dim,ub,lb)
Boundary_no= size(ub,2); % numnber of boundaries
% If the boundaries of all variables are equal and user enter
    a signle
% number for both ub and lb
if Boundary_no==1
    X=rand(SearchAgents_no,dim).*(ub-lb)+lb;
end
% If each variable has a different lb and ub
if Boundary_no>1
    for i=1:dim
        ub_i=ub(i);
        lb_i=lb(i);
        X(:,i)=rand(SearchAgents_no,1).*(ub_i-lb_i)+lb_i;
    end
end


function [RWs]=Random_walk_around_antlion(Dim,max_iter,lb,
    ub,antlion,current_iter)
if size(lb,1) ==1 && size(lb,2)==1 %Check if the bounds are
    scalar
    lb=ones(1,Dim)*lb;
    ub=ones(1,Dim)*ub;
end
```

```
if size(lb,1) > size(lb,2) %Check if boundary vectors are
   horizontal or vertical
    lb=lb';
    ub=ub';
end
I=1; % I is the ratio in Equations (2.10) and (2.11)
if current_iter>max_iter/10
    I=1+100*(current_iter/max_iter);
end
if current_iter>max_iter/2
    I=1+1000*(current_iter/max_iter);
end
if current_iter>max_iter*(3/4)
    I=1+10000*(current_iter/max_iter);
end
if current_iter>max_iter*(0.9)
    I=1+100000*(current_iter/max_iter);
end
if current_iter>max_iter*(0.95)
    I=1+1000000*(current_iter/max_iter);
end
% Dicrease boundaries to converge towards antlion
lb=lb/(I); % Equation (2.10) in the paper
ub=ub/(I); % Equation (2.11) in the paper
% Move the interval of [lb ub] around the antlion [lb+anlion
   ub+antlion]
if rand<0.5
    lb=lb+antlion; % Equation (2.8) in the paper
else
    lb=-lb+antlion;
end
if rand>=0.5
    ub=ub+antlion; % Equation (2.9) in the paper
else
    ub=-ub+antlion;
end
% This function creates n random walks and normalize accroding
   to lb and ub
% vectors
for i=1:Dim
    X = [0 cumsum(2*(rand(max_iter,1)>0.5)-1)']; % Equation
       (2.1) in the paper
    %[a b]--->[c d]
    a=min(X);
    b=max(X);
    c=lb(i);
    d=ub(i);
    X_norm=((X-a).*(d-c))./(b-a)+c; % Equation (2.7) in the paper
    RWs(:,i)=X_norm;
End
```

```
function choice = RouletteWheelSelection(weights)
  accumulation = cumsum(weights);
  p = rand() * accumulation(end);
  chosen_index = -1;
  for index = 1 : length(accumulation)
    if (accumulation(index) > p)
      chosen_index = index;
      break;
    end
  end
  choice = chosen_index;
```

Index

Printed and bound by CPI Group (UK) Ltd, Croydon, CR0 4YY

24/10/2024

01778306-0002